Green Lighting

Other McGraw-Hill Books of Interest

Build Your Own Electric Bicycle by Matthew Slinn

Build Your Own Electric Motorcycle by Carl Vogel

Build Your Own Electric Vehicle, Second Edition by Seth Leitman and Bob Brant

Build Your Own Plug-In Hybrid Electric Vehicle by Seth Leitman

Do-It-Yourself Home Energy Audits: 140 Simple Solutions to Lower Energy Costs, Increase Your Home's Efficiency, and Save the Environment by David S. Findley

Fuel Cell Projects for the Evil Genius by Gavin Harper

Green Projects for the Evil Genius by Jamil Shariff

Recycling Projects for the Evil Genius by Russel Gehrke

Renewable Energies for Your Home: Real-World Solutions for Green Conversions by Russel Gehrke

Run Your Diesel Vehicle on Biofuels: A Do-It-Yourself Manual by Jon Starbuck and Gain Harper

Solar Energy Projects for the Evil Genius by Gavin Harper

Solar Power for Your Home: Step-by-Step Plans for Do-It-Yourself Solar Projects by David S. Findley

Green Lighting

How Energy-Efficient Lighting
Can Save You Energy and Money and
Reduce Your Carbon Footprint

Brian Clark Howard
William J. Brinsky
Seth Leitman

Mc
Graw
Hill

New York   Chicago   San Francisco
Lisbon   London   Madrid   Mexico City
Milan   New Delhi   San Juan
Seoul   Singapore   Sydney   Toronto

![The McGraw-Hill Companies]

Library of Congress Cataloging-in-Publication Data

Howard, Brian Clark
   Green lighting : how energy-efficient lighting can save you energy and money and reduce your carbon footprint / Brian Clark Howard, William J. Brinsky, Seth Leitman.
      p. cm.
   Includes index.
   ISBN 978-0-07-163016-0 (alk. paper)
   1. Electric lighting—Energy consumption. 2. Energy conservation. I. Brinsky, William J. II. Leitman, Seth. III. Title.
   TK4188.H69   2011
   644'.3—dc22

2010017160

McGraw-Hill books are available at special quantity discounts to use as premiums and sales promotions, or for use in corporate training programs. To contact a representative please e-mail us at bulksales@mcgraw-hill.com.

Green Lighting

1 2 3 4 5 6 7 8 9 0 DOC/DOC 1 9 8 7 6 5 4 3 2 1 0

ISBN 978-0-07-163016-0
MHID 0-07-163016-3

 The pages within this book were printed on acid-free paper containing 100% postconsumer fiber.

Sponsoring Editor	**Proofreader**
Judy Bass	Paul Tyler
Acquisitions Coordinator	**Indexer**
Michael Mulcahy	Karin Arrigoni
Editorial Supervisor	**Production Supervisor**
David E. Fogarty	Pamela A. Pelton
Project Manager	**Composition**
Patricia Wallenburg	TypeWriting
Copy Editor	**Art Director, Cover**
James Madru	Jeff Weeks

About the Authors

Brian Clark Howard (New York, NY) is a Web editor at *The Daily Green*, which is part of Hearst Digital Media and is one of the world's largest and most trusted sources for consumer information on living a more environmentally friendly life. Brian was previously managing editor of *E/The Environmental Magazine*, the oldest and largest independent environmental magazine in the United States. He has written for Yahoo!, MSN, *Plenty*, *The Green Guide*, *Popular Mechanics* online, *Men's Health*, *Mother Nature Network*, *Oceana*, AlterNet and elsewhere. Brian coauthored a book on geothermal heating and cooling, forthcoming from McGraw-Hill, and the chapter on saving energy for the 2009 book *Whole Green Catalog* (Rodale). Brian earned an MS in journalism from Columbia University and holds two bachelor's degrees in environmental sciences. Brian was a finalist for the 2005 Reuters/IUCN Environmental Media Awards and has appeared on numerous radio and television programs. He blogs for Asylum and as the URTH Guy at *The Daily Green*.

Bill Brinksy (Harriman, NY) worked for Con Edison for 10 years in billing. Bill also was managing partner of Project Management Consultants before founding Envirolite Systems in 1994. Envirolite focuses on lighting and energy, from design of new systems to energy-efficient upgrades of current systems.

Seth Leitman (Briarcliff Manor, NY) is currently president and managing member of the ETS Energy Store, LLC (www.etsenergy.com), which sells organic, natural, and sustainable products for business and home use (from energy-efficient bulbs to electric vehicle conversion referrals). Previously, he worked for the New York State Power Authority and the New York State Energy Research and Development Authority, where he helped develop, market, and manage electric and hybrid vehicle programs serving New York State and the New York metropolitan area. Seth wrote the book *Build Your Own Electric Vehicle*, Second Edition, with Bob Brant. He also writes for Greenopia and Planet Green online and runs a blog called www.greenlivingguy.com, which discusses sustainable living, energy efficiency, and electric cars.

APR 19 2011

Contents

Preface

How we light up the places we live and work in makes a big impact on how we feel, as well as how we accomplish our daily tasks. It also has a tremendous impact on the environment. The kinds of bulbs and fixtures we use, along with the habits we keep, matter, probably in a number of ways you've never thought of. We begin with the simple fact that a conventional incandescent bulb—the standard light bulb—turns only 2 to 10 percent of its consumed energy into light, whereas the rest goes out as wasted heat. From there, there's practically no limit to how green your lighting can become. Read on to find out how.

Acknowledgments

Brian: I would like to thank Judy Bass, Patricia Wallenburg, and the McGraw-Hill team for their professionalism and support of such important projects. Thank you also to Seth Leitman, who honored me with the invitation to help with this book. I would like to thank my mentors, Doug Moss and Jim Motavalli, who taught me so much about the possibilities of going green and life at *E/The Environmental Magazine*. As Jim told me when I started as an intern, it really is possible to change the world. I also need to thank my brilliant colleagues at *The Daily Green*, Dan Shapley and Gloria Dawson, who teach me and inspire me every day. I want to thank Remy Chevalier, who has taken many hours to explain complex topics to me and who has steered me to many invaluable sources. Remy also helped to prepare the resource guide at the end of this book. Check out his outstanding environmental library in Connecticut (The Aquarium), and find him on remyc.com, where he promotes green lighting as one of the brightest ideas for a better world. I am indebted to David Bergman for reading this manuscript and sharing his expert experience. Find him at www.cyberg.com. I'd also like to thank all my friends in the green blogosphere and throughout the green movement. There are too many names to list here, which perhaps is a testament to how collaborative, supportive, and creative this space is. Every day I am honored to be a part of it.

Last but not least, I'd like to thank my family, who taught me to respect and appreciate the natural world and to strive to leave everything better than how I found it. Thanks to my parents, Allan and Diana, and my sisters, Amy and Lisa. Thank you also to my wonderful, beautiful girlfriend, Gloria, who supports and challenges me.

Bill: I would like to thank my mother, for always encouraging my curiosity.

Seth: To paraphrase an actor who just won an Emmy, "There are so many people to thank." However, I first want to dedicate this book to my family, to my beautiful wife, Jessica, and my beautiful sons, Tyler and Cameron. I hope that this book helps in some way to create a better world as they grow up. They have watched my involvement in electric cars expand over the years, and with global warming looming large and green becoming the new black, they really appreciate what I am doing.

I also want to thank Judy Bass from McGraw-Hill, who believed in me. I have always thought that people come into your life for a reason, and she blessed me with the opportunity to write this book and give this industry the jolt it needs. She believed in preparing a book with mass appeal while keeping a technical approach.

CHAPTER 1

The Benefits of Changing Your Lighting

Unless you work in film, photography, or interior design, chances are good that you haven't given a whole lot of thought to lighting. Yet lighting is a constant component of our home and work environments and affects our lives in many different ways. Beyond helping us to see and complete our daily tasks, lighting affects how we act and feel.

Lighting can play an essential role in how we design our homes. Since lighting is also a major contributor to global energy consumption and pollution (not just through the energy it uses but also through the hardware itself), we have strong incentives to seek the most sustainable, environmentally friendly options possible. That is *green lighting* (Figure 1-1).

This book will identify a number of aspects of sustainable lighting and show you how you can save money, live and work smarter, and improve your decor, all while helping the planet. Key topics include:

- Why we should green our lighting
- What lighting products are the most effective at saving energy and are made with sustainable materials
- Why certain lighting technologies are better than others and in what circumstances
- How to install these technologies in your home or business
- How to get the most out of lighting in your life

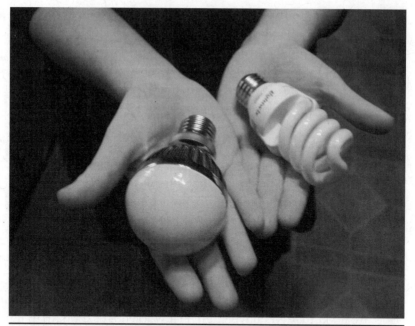

Figure 1-1   Light-emitting diode (LED) bulbs (*left*) are gradually replacing compact fluorescent lights (CFLs) in greener lighting because they are more efficient and last longer. *(Photo by Brian Clark Howard)*

- Where to obtain green lighting technologies
- What advanced lighting technologies are in development and may be available soon

Lighting's Big Footprint

Americans spend, on average, 9 to 20 percent of our home electric bills on lighting, according to the U.S. Department of Energy (DOE) (see Figure 1-2). This amounts to an average of $90 to $180 a year because our average annual electric bill is $900 a year.

For commercial buildings, lighting is responsible for an average of 38 percent of electricity use (see Figure 1-3). This is especially significant because energy costs for commercial buildings typically are 30 percent of their total operating expenses.

As a nation, we spend about one-quarter of all electricity we use on lighting, at a cost of $37 billion annually. The buildings sector as a whole is responsible for approximately 66 percent of U.S. electricity use. Office buildings alone consume 23 million megawatt-hours

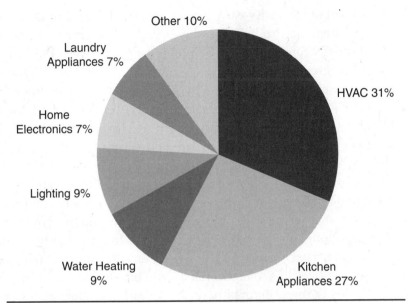

FIGURE 1-2   A breakdown of the average annual electricity use in an American home. *(Pew Center on Global Climate Change)*

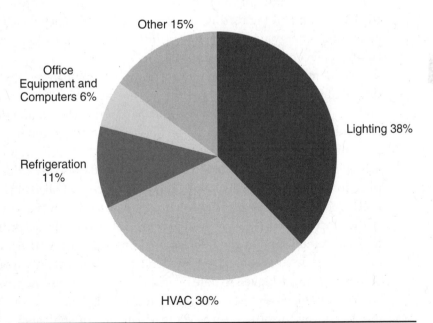

FIGURE 1-3   A breakdown of the average annual electricity use in an American commercial facility. *(Pew Center on Global Climate Change)*

(MWh) of electricity each year, or 28 percent of all commercial energy demand. An important, and perhaps alarming, point is that electricity consumption in buildings *doubled* between 1989 and 2005. If this growth rate continues, electricity demand in buildings will increase another 150 percent by 2030, according to Flex Your Power, California's state marketing program for energy efficiency. Globally, about 20 percent of electricity is dedicated to lighting, which is more than what is produced by hydro- or nuclear power.

Americans spend approximately $11.9 billion a year on lighting equipment—$2.7 billion of this on lamps, with the remainder divided among fixtures, components (including ballasts and controls), and associated services, according to a report for the DOE by Navigant Consulting, Inc. Globally, the market for lighting hardware is estimated to be more than $100 billion a year, with lamps making up $20 billion, according to Freedonia Group, Inc.

While these figures may seem daunting, they also hint at a tremendous opportunity to enact savings and pass the benefits to shareholders, customers, constituents, and home owners. Roughly two-thirds of all the electricity we use in the United States is generated through burning fossil fuels, so decreasing the amount we consume can have an enormous environmental impact. According to the DOE, undertaking readily available energy-efficiency measures can reduce energy use—and thus utility bills—by up to 65 percent (with 30 percent being common) (Figure 1-4).

For commercial buildings, retrofitting lighting can readily reduce that part of the bill by 40 percent, according to Flex Your Power, and the changes can be implemented as part of routine maintenance schedules. True, current lighting technology can be as much as 98 percent more efficient than standard bulbs and fixtures, although achieving such a high reduction across the board, in real-world applications, isn't yet practical. Still, about 10 percent of all electricity used globally could be saved if we did manage to switch everything to the most efficient lighting systems currently available, according to a report published by the International Energy Agency (IEA). The carbon dioxide emissions saved by such a transition certainly would dwarf anything achieved to date by installing wind and solar power.

What perhaps is most exciting is that it doesn't take very long for green lighting to show returns on investment. An analysis by the Energy Cost Savings Council estimates the average payback period

FIGURE 1-4   About a quarter of electricity use in the United States goes to lighting, at a cost of $37 billion annually. *(Photo by Brian Clark Howard)*

for lighting retrofits as 2.2 years, with an average return on investment (ROI) of 45 percent. According to the report, lighting upgrades are second only to energy meters and monitors in rapid payback and are significantly better than on-site power generation, which takes an average of 4.3 years.

Companies See the Light—and the Savings

The U.S.'s largest real estate investment trust, Equity Office Properties, recently conducted a retrofit of lighting systems in 20 of its buildings, and this resulted in savings of $1.7 million. At its New Jersey facility, pharmaceutical maker Warner-Lambert (now part of Pfizer) replaced more than 100,000 fluorescent and incandescent lamps with more efficient bulbs, in addition to taking other energy-efficiency steps. The upgrade saves $1.4 million per year in energy costs and reduced carbon dioxide emissions from electricity use by 11,000 tons per year. La Quinta hotels recently installed energy-efficient lighting in many of its properties, reducing overall electricity use by eight percent, saving $1.3 million in annual energy costs, according to an analysis by Innovest Strategic Value Advisors.

Another company that is making incremental changes to something we stare at every day is Starbucks. The coffee giant has been replacing traditional lights in thousands of its shops with superefficient LEDs, which use considerably less energy than even compact fluorescent lighting (CFL). This will have a long-term direct impact on the company's bottom line in part because the LEDs are expected to last 10 years and therefore save on replacement costs, reduce waste, and boost organizational efficiency (Figure 1-5).

A chain of 47 groceries across West Texas, United Supermarkets, recently changed 3,600 of its refrigerator and freezer lights from fluorescents to LEDs. The company expects to save more than $633,000 annually as a result ($369,000 in energy savings and $264,000 in maintenance savings). The expected ROI is a brief 1.8 years. The LEDs were made by General Electric (GE) subsidiary Lumination, LLC, and are rated for 50,000 hours. They provide uniform color and up to three times the light-level uniformity of the previous fluorescent lamps. The estimated environmental impact is an annual 2.9 million pound reduction in carbon dioxide emissions, equivalent to planting 364 acres of trees or removing about 257 cars from the road.

"Lowering our energy costs is very important to us," Michael Molina, vice president of facilities and design at United Supermarkets, explained. "Doing all we can to be environmentally conscious is also important. We are a 92-year-old company now in its third generation of family ownership, and we want to make our operations better now and for the next generation."

On a recent tour of Hilton New York in midtown Manhattan, the hotel's director of sustainability (and assistant director of food and beverage), Jason Tresh, explained how the largest hotel in New York City has been decreasing its lighting bills. This is no small payback for a facility with more than 2,000 guest rooms, plus restaurants, shops, ballrooms, and more. "Of course, we're in business to make money, but we are also trying to do it sustainably. At the end of the day, we feel good and our guests appreciate it," said Tresh, who pointed out that the flagship hotel is serving as a learning example for the global company (major green initiatives are also under way at Hiltons in San Francisco and Baltimore, the latter of which even has a living green roof).

FIGURE 1-5 LED track lights can be seen in this model "green" Starbucks shop in New York City. The company is installing efficient lighting in thousands of its locations. *(Photo by Brian Clark Howard)*

Tresh introduced Hilton New York's chief electrician, Rick Gonzalez, who explained that 90 percent of the massive building is now lit with fluorescents and LEDs. "We stepped down from T12 to T8 and T5 fluorescents to save energy and have been introducing some LEDs," Gonzalez explained (see Chapter 4 for a discussion of fluorescent ratings). "We have switched from magnetic to electronic ballasts [which are more efficient], public areas are dimmed down to

20 percent at night, and we put occupancy sensors in meeting rooms," he added. The only incandescents remaining are in the ballrooms, where Gonzalez said he needed to preserve the traditional warm feel of the light. Perhaps not surprisingly given the contagious nature of going green, the hotel also has gone beyond lighting to make other changes, including adding a fuel cell on the roof for clean power, robust recycling, low-flow water fixtures, and a food waste digester.

Lighting a Cozy, Functional Home—for Less

It's important to remember that the benefits of green lighting aren't limited to commercial settings: Home owners also can reap the rewards of a brighter future! When the Lancaster-Hageman family built a gorgeous dream house on the coast in Narragansett, Rhode Island, in 2009, the family included advanced lighting features that enhance comfort and beauty, as well as slash energy consumption. In an in-person interview, Kim Lancaster explained that she had been inspired by her kids (ages three and six) and what she had been learning about the environment and their health.

"It started with a chemical-free mattress," Lancaster said of her journey to design a home that would be safe, nurturing, and comfortable. The home is certified Gold through Leadership in Energy and Environmental Design (LEED) and is Energy Star rated (learn more about these programs in Chapter 2). It is 42 percent more energy efficient than traditional homes, yet it is packed with advanced technology, such as a home video server, a whole-house music system, integrated security, and wiring for solar panels, which the family may invest in down the line (Figure 1-6).

Lancaster said that she isn't crazy about CFLs, so she used them sparingly, opting instead for high-efficiency LEDs for most fixtures, both indoors and outdoors. "The LED lights were really expensive, but I think it was worth it," she said. "Lighting is only 8 percent of our bill, which is really low, and [the LEDs] produce a really warm light; it's not blue at all." (Of course, as technology improves, the costs of LEDs should come down dramatically. Learn all about them in Chapter 5.)

Lancaster is particularly fond of the Juno LED recessed lights and says that she can easily light specific areas to set scenes. With the

FIGURE 1-6   The home Kim Lancaster and Joe Hageman built in Rhode Island is a gorgeous showcase of green lighting technology. *(Ashley Daigneault/Caster Communications)*

flip of a switch, she or her husband can illuminate a nighttime path from the master bedroom to the kids' rooms. They also programmed their LED holiday lights on a timer so that they never had to remember to turn them on or off. All lighting is tied to the Control4 whole-house management system, which makes it easy to optimize and schedule heating and cooling, security, and entertainment (more on controls in Chapter 6). They have on/off buttons in their cars; a "home" button that turns on key lights, such as the entryway and kitchen, with a single switch; and an on/off button beside the master bed, which can power up or down all lighting in the house. Everything is dimmable, and the family set the default brightness at 85 percent to save even more.

"I started out of frustration as a home owner who wanted a green home, with some luxury elements, and a lot of technology in it," said Lancaster. "People said it couldn't have technology and be green, so I wanted to prove them wrong. You don't have to give up everything to go green."

The family's system is a good example of some of the functional benefits of new green lighting technology. As Jay McLellan, president and CEO of Home Automation, Inc., recently explained at the

Greener Gadgets 2010 conference in New York, "If you decrease the comfort of a home, you're going to get put out on the back porch with the milk bottles." McLellan's firm works with upscale customers in New Orleans to design integrated energy and technology management systems that provide convenience, customization, and energy savings.

Kim Lancaster and Joe Hageman built their new home as a showpiece for green design, and they blog about it at their Web site Green Life Smart Life (greenlifesmartlife.com). They also had several corporate sponsors, including Lutron, Kohler, and the Consumer Electronics Association. But you don't have to work in public relations, as Lancaster does, or dive headfirst into a deep green pool in order to save some energy. You can start by dipping in a toe by buying just one new light bulb.

Ask yourself: *How would you like to pay an electric bill that's nearly a third less than last year?*

To picture this on a large scale, a 30 percent reduction in energy consumption can lower operating costs by $25,000 per year for every 50,000 square feet of office space. In fact, according to research conducted by the Environmental Protection Agency (EPA), for every $1 invested in energy efficiency, asset value increases by an estimated $3. This, in turn, can help facility managers to get more favorable loan conditions. It means that improving efficiency in lighting can be a low-risk way to boost the bottom line over time. And whether you are a homeowner or the manager of a large facility, much of these savings can be realized by no-cost or low-cost projects, which we will show you in this book.

Riding the Wave of Green Awareness

We remember back in the early 1990s when lawmakers and stakeholders were meeting in New York State to talk about developing some green building tax credits, and educated people wondered if that meant simply changing the color of the paint on the building. Now people are much more aware about sustainable design and conservation and about the importance of reducing fossil fuel use and global warming emissions. Energy efficiency is the quickest and easiest way for society to start going green and achieving the goals of real energy independence and clean air and water.

We have seen many companies and public institutions save millions of dollars in energy costs and prevent millions of tons of emissions by switching to more energy-efficient lighting. And as we become even more sustainable as a society, greening our lighting will get that much easier. More advanced systems will become more widely used and will provide brighter light of higher quality, color, and versatility—at much reduced prices. Soon, more manufacturers will take a more holistic approach to lighting. Fixture materials will have more recycled content, toxins will be eliminated, and materials will be sourced locally, further reducing emissions by decreasing transportation needs.

Specific Benefits of Green Lighting

The most common complaints we hear about green lighting are that the technology is too expensive and simply "not feasible." In our experience, though, no lighting retrofit has ever cost more than the long-term savings it provides. Note that this isn't necessarily true with some environmentally friendly technologies, such as some advanced alternative-energy projects. In addition to slashing energy costs, switching to greener lighting can provide numerous other benefits, including the following.

Better Overall Economic Health

It's no surprise that an efficiently run household is often a productive and comfortable place, and the same holds true for larger organizations. This is perhaps why a 2002 analysis by Innovest Strategic Value Advisors linked improving energy efficiency of companies to better stock market performance. It seems that analysts and investors are starting to recognize that management teams that are adept at trimming energy expenses also tend to be good at steering the rest of the business. Plus, more and more businesses are being asked to disclose environmental and energy-performance data as part of their annual reports. Showing improvement can help them to win business from government agencies and other organizations that are increasingly interested in environmental responsibility.

FIGURE 1-7   CFLs (*left*) and LEDs (*right*) produce increasingly high-quality light. (*Photo by Brian Clark Howard*)

Efficiency upgrades can help a company to get on the radar of *socially responsible investing* (SRI) mutual funds and analysts, thus opening up new avenues of business. SRI funds screen out companies that managers deem unsavory and invest in firms that are leading the way in sustainability and social values. Moreover, the positive karma—and publicity—organizations earn for doing better by the environment can be invaluable. Improving lighting efficiency typically is the lowest-hanging fruit and often yields the quickest payoff (Figure 1-7).

Improved Property Values

An increasing number of home builders and real estate agents are discovering that they can get a leg up in the marketplace by advertising that a property has green features—and once again, green lighting is often the easiest place to start. Walter Molony, a spokesperson for the National Association of Realtors, told *The Daily Green*, "People definitely value energy efficiency. As utility costs continue to rise, it becomes greater value in people's minds."

The number of homes that were built according to voluntary green building standards ballooned by 50 percent between 2004 and 2007, reports the National Association of Home Builders (NAHB). Although new home starts have dropped dramatically in the wake of the housing market crash and global recession, interest in greener features remains high. More than half the NAHB's 235,000 members (representing about 80 percent of U.S. home builders) reported in 2007 that they were starting to use at least some green building practices.

Further, real estate professionals are starting to notice that owners of green homes tend to be happier than when they live in more conventional digs, as reported by a recent NAHB/McGraw-Hill Construction survey. It's worth noting that almost 40 percent of Americans who recently renovated their dwellings did so with at least some green products.

In a troubled market, being able to concretely show potential buyers that they will save money on their utility bills, as well as live smarter and more comfortably, can help to make your property more attractive than the competition. When the market is hot, having green features is like icing on the cake, and it can help you to access the growing segment of consumers who are deeply concerned about environmental issues.

The same holds true on the commercial side. In fact, a study by the Institute for Market Transformation found that $1 invested in energy efficiency with a 20 percent ROI could increase commercial property value by $2.

Improved Comfort, Employee Attendance, and Tenant Retention

It is perhaps not surprising that most energy-efficiency measures also improve the comfort and attractiveness of the indoor environment. Well-designed lighting retrofits, while reducing energy consumption, also improve *visual acuity*—the ability to see details well. Better vision, in turn, helps workers to complete tasks faster and reduces eye and mental strain. Better mood lighting helps to foster a pleasant environment that can bring out the best in staff and visitors.

A study by the Rocky Mountain Institute, a progressive energy think tank in Colorado, found that high-efficiency lighting with improved light quality, intensity, and color dramatically reduced worker eye strain, vision-related errors, and even absenteeism. A

FIGURE 1-8 This experimental "green" Walmart superstore in Aurora, Colorado makes extensive use of daylighting, as well as efficient lighting technologies. *(Photo by Brian Clark Howard)*

number of occupational analysts have pointed out that workers who get fatigued less quickly are less likely to call in sick or get in on-site accidents. And those who are more comfortable in their workplace are more likely to stay with the company (Figure 1-8).

For income-generating properties, better lighting can help improve tenant recruitment and retention. For home owners, better lighting means more personal comfort and better utility from being inside, which is significant because studies show that we are indoors for up to 90 percent of our time these days. Don't underestimate how much interior lighting affects the way we feel and the way we work and live.

Lockheed Martin Space Systems recently built 33,000 square feet of facilities in Sunnyvale, California, that meet the rigorous LEED standards maintained by the U.S. Green Building Council (USGBC). The many green features include efficient lighting, full-cutoff fixtures to reduce light pollution outside, and promotion of *daylighting* through windows and sloped ceilings. Managers told the USGBC that after they moved into the new site, they saw a 15 percent drop in employee absenteeism. This resulted in savings that they said made up for the building's green cost premium in the first year alone. Incidentally, that additional cost for installing green features was only 1.5 percent above conventional construction, according to Lockheed, further dispelling the belief that going green has to cost a lot of money.

"When they're designed well, green buildings are very competitive on initial cost, and they have lower operating costs by using less energy and water," Kaushik Amruthur, a Lockheed senior facilities engineer, explained.

Increased Productivity

A number of studies from the Rocky Mountain Institute and others have correlated increased worker productivity with better and more efficient lighting. Nationally, improvements to indoor environmental conditions are estimated to have generated $20 billion to $160 billion from greater workforce productivity, according to a report in the July-August 2002 *Annual Review of Energy and the Environment.*

In a 1986 effort that is often cited by supporters of green building, the Main Post Office in Reno, Nevada, installed an efficient lighting system and lowered the ceiling, which made the room easier to heat and cool. Harsh direct downlighting was replaced with indirect lighting using long-lasting bulbs. After 40 weeks, worker productivity reportedly had increased by more than eight percent.

In some cases, improving the lighting of physical work environments also may help to attract the best and the brightest workers. This can lead to significant increases in business, and it is money up for grabs. As Kaushik Amruthur of Lockheed put it, today's employees are more discriminating about the environmental qualities of the buildings in which they work. If workers slave away in dark and dingy conditions, they are more likely to feel undervalued by their employers and therefore less interested in going the extra mile.

These benefits also extend to schools, although an estimated 40 percent of American schools suffer from poor environmental conditions that can compromise the health and learning of students, according to the USGBC. However, a 2005 study in Washington State by Paladino & Company found a 15 percent reduction in student absenteeism at green schools (Figure 1-9). A 2006 Capital E review of 30 green schools across the country concluded that "based on a very substantial data set on productivity and test performance of healthier, more comfortable study and learning environments, a three to five

Figure 1-9   Superefficient LEDs are showing up in an increasing range of applications. *(Photo by Brian Clark Howard)*

percent improvement in learning ability and test scores in green schools appears reasonable and conservative."

Boosting Sales

It is perhaps common sense that well-lit stores see more foot traffic and better sales than dim shops. Lighting makes a difference. At a recent visit to the Consumers Union Laboratories in New York state, the tech reviewers for *Consumer Reports* magazine explained that TV showrooms jack up the brightness and contrast on sets in order to catch the eyes of shoppers and move more units. Better news for greens is that daylighting can also raise sales. In 1999, the Heschong Mahone Group surveyed 108 outlet stores operated by the same chain, and found that sales increased by 40 percent in stores that had installed skylights.

Michael A. Steele, chief operating officer (COO) of Equity Office Properties, told Innovest Strategic Value Advisors that "having an energy-efficient building gives the owner an opportunity to win over the customer, especially when that is the difference between otherwise similar buildings."

Lighting and the Environment

Most lighting is powered by our electricity grid, which, in turn, is supplied largely through burning fossil fuels. Not only are fossil fuels a finite resource that are difficult and costly to extract, but they release emissions that contribute to global climate change, cause air-quality issues such as acid rain and smog, and pose significant risks to human health. It isn't often reported in the media, but three million people die each year around the planet from the effects of air pollution, according to the World Health Organization. To put this into some perspective, this is three times the number who die each year in car accidents. In the United States, traffic fatalities total roughly 40,000 per year, whereas air pollution claims an estimated 70,000 American lives annually. U.S. air pollution deaths are equal to deaths from breast cancer and prostate cancer combined.

You may have heard the TV commercials that point out that if every American replaced just one regular incandescent light bulb with a CFL, we would save enough energy to light more than three million homes and prevent greenhouse gas emissions equivalent to that of 800,000 cars. In fact, in 2007 alone, Americans saved $1.5 billion by switching to CFLs, enough power to light all the households in Washington, DC, for 30 years. This is equivalent, in terms of reducing greenhouse gases, to planting 2.85 million acres of trees or taking two million cars off the road each year. If everyone replaced their five most frequently used bulbs with CFLs, we would save $8 billion a year in energy costs.

If we went further and more completely improved our lighting efficiency by upgrading switches, controls, wires, and fixtures, we would have the potential to reduce greenhouse gases equivalent to the emissions from nearly *20 million cars*, according to the DOE. Studies have suggested that a complete conversion to green lighting could decrease CO_2 emissions from electric power use for lighting by up to 50 percent in just over 20 years in the United States.

While CFLs have been seen as an exciting greener technology, many people believe that LEDs, and the devices that follow them, hold the best potential for the future. A recent report by McKinsey & Company cited conversion to LED lighting as potentially the most cost-effective approach, out of a range of possibilities, to tackle global warming using existing technology. Activist Remy Chevalier of Connecticut-based Rock The Reactors argues that a broad switch to LEDs would save so much power that we could shut down our aging—and controversial—nuclear power plants.

Green Lighting Is Becoming Law

In June 2009, President Barack Obama stood with Energy Secretary Steven Chu in the grand foyer of the White House to announce new lighting efficiency standards. "Now, I know light bulbs may not seem sexy, but this simple action holds enormous promise," Obama said. "And, by the way, we're going to start here at the White House. Secretary Chu has already started to take a look at our light bulbs, and we're going to see what we need to replace them with energy-efficient light bulbs."

The two Nobel Prize winners estimated that between 2012 (when the rules start to take effect) and 2042, the new standards will save consumers up to $4 billion a year. In fact, Congress had already passed a law—the Energy Independence and Security Act of 2007—that mandates the phaseout of inefficient bulbs (effectively, typical incandescents). Specifically, all general-purpose light bulbs that produce 310 to 2,600 lumens of light must be 30 percent more energy efficient than current incandescent bulbs by 2012 to 2014. The efficiency standards start with 100-watt bulbs in January 2012 and end with 40-watt bulbs in January 2014. Certain specialty bulbs are exempt. By 2020, a second tier of restrictions would kick in, requiring all general-purpose bulbs to produce at least 45 lumens per watt (about the same as current CFLs). Even then, some products will be exempt, including floodlights, three-way bulbs, candelabras, and colored bulbs.

Many other countries around the world are getting serious about green lighting and are passing laws that reduce the allowable wattages of bulbs. Australia became the first country to announce an outright ban on incandescent bulbs, to take effect in 2010. The country's environment minister, Malcolm Turnbull, said that by banning incandescent bulbs, Australia should reduce annual carbon dioxide emissions by four million tons. Brazil and Venezuela started to phase out incandescents back in 2005 and have since been joined by Argentina. Russia and the European Union are currently working on phaseouts, and Canada plans to restrict the bulbs by 2012.

Toshiba recently announced that it will cease production of incandescent bulbs, after having shipped more than four billion of the products over the last 120 years. Toshiba made seven million incandescents last year and 14 million CFLs, according to Mis-Asia and Treehugger.com. The company's decision to focus on greener lighting, including LEDs, is estimated to cut 430,000 tons of CO_2 emissions per year once the products enter the market.

Summary

As you now know, greening your lighting has many benefits, whether you manage a big commercial facility or live in a one-room apartment. As you can see in Figure 1-10, greener bulbs pay for themselves in a short time, and this book will show you lots of ways to save even more energy.

We recently got a chance to see the exciting new LED technology that ANL (Andy Neal Lighting) has been installing on the Poughkeepsie-Highland Railroad Bridge in upstate New York. It's an ambitious project, illuminating the world's tallest pedestrian bridge, which soars 212 feet above the majestic Hudson River. The 120-year-old railroad crossing has been reenvisioned as Walkway Over the Hudson State Park, and it was a centerpiece of the celebration of the

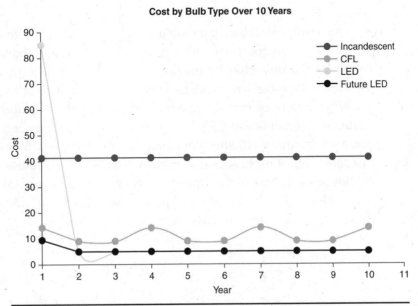

FIGURE 1-10   Over a fairly short time, greener bulbs start saving money. *(Courtesy of Wikimedia Commons)*

four-hundredth anniversary of Henry Hudson's historic sail up the river. The 6,767-foot-long bridge is emblematic of a link to the past and the quest for a more sustainable future, much like green lighting. Andy Neal's LEDs are brilliantly bright, perhaps symbolizing glowing promise of what's to come.

Tomorrow's LEDs and other advanced green lighting technology will be even more adaptable, cheaper, and more efficient. It also increasingly will be made out of recycled, nontoxic, and locally sourced materials. In order to get there, though, we need to support the green lighting technology of today, which will drive further demand, support research and development, and give us many benefits in the short term.

So what have we learned?

- We have seen that when you upgrade to green lighting, you will reduce energy costs and pay yourself back in relatively "no time" (Figure 1-10).
- We have seen that improving lighting can provide many other benefits, including enhanced comfort, reduced absenteeism, higher productivity, and higher property values.
- We have seen that green lighting reduces greenhouse gases and pollution.

Even if everyone takes it one bulb at a time, we can all switch to green lighting. And we don't have to sacrifice style or design, thanks to great new fixtures. Plus, by making smart use of daylighting and solar lamps or tubes, we won't need the electrical grid at all!

When one of us started greening his own lighting at home, it started with one 14-watt CFL, bought from a store. However, when you see that utility bill after your first month of savings, $5 in this case, and you count how many bulbs you have in your house, you quickly grasp the potential. Then you want to do more, so you buy two or three more bulbs. Maybe you rewire a "high-hat" recessed fixture. Now you are on a mission!

CHAPTER 2

Basic Lighting Terms and Explanations

What Is Efficient Lighting?

Now that you know how important it is to save energy in lighting, as well as how valuable it can be to home owners and property managers, you're probably looking forward to getting started. So what exactly is efficient lighting? In the simplest terms, to start saving energy, you must either reduce the amount of time a bulb is on or reduce the amount of electricity the bulb consumes. Let's break these concepts down:

- *Reduce the amount of time lights are used.* Turn off lights you aren't using! Use task lighting and daylighting (through windows and skylights) instead of big overheads. Install, and properly use, timers, occupancy sensors, and other controls. We'll go over all these areas in more detail in the coming pages.
- *Lower the bulb wattage.* Replace bulbs and/or fixtures with units that provide an equivalent amount of light but require less energy. This can be done with dimmers and by replacing inefficient incandescent bulbs with the following common types of lights:
  - *Halogen bulbs* (roughly 30 percent more efficient). Halogens are incandescents that are enhanced with some extra technology, including the presence of a halogen (a nonmetal element from Group 17 of the periodic table, either fluorine, chlorine, bromine, iodine, or astatine). They have some drawbacks, however, including high heat production and higher cost.

FIGURE 2-1 Better lighting stores offer several different technologies, and it's always a good idea to try before you buy. *(Photo by Brian Clark Howard)*

- *Compact fluorescent light (CFL) bulbs* (75 percent more efficient). Better than halogens are CFLs, which last longer and have come a long way in quality in recent years. CFLs do contain trace amounts of mercury, so they must be handled carefully, and they don't work well for all settings.
- *Light-emitting diodes* (LEDs) (up to 90 percent more efficient) *or other emerging advanced technology.* LEDs benefit from ruggedness and very long life, and they don't give off much heat. They are still relatively expensive, although prices have been falling.

In order to get the most out of an energy-efficient lighting upgrade, it's helpful to know a bit about light, electricity, and utilities.

What Is Light?

Light is all around us but is usually taken for granted, unless you find yourself lost in a cave or trying to get grass to grow in a shady spot on your lawn. Scientifically, light is usually described as the type of electromagnetic radiation that has a wavelength visible to the human eye (roughly 400 to 700 nanometers). Light exists as tiny "packets" called *photons* and exhibits the properties of both particles and waves. Don't worry if you don't really understand what this means; it's something the world's top physicists have debated for many years.

What Is Lighting?

There are many different types of lighting, most of which you will learn about in this book. All lights have a *lamp*, which is commonly referred to as a *light bulb*. The lamp is connected to the energy source by something called a *fixture*, which also positions the light in a useful way. In fluorescent and high-intensity-discharge (HID) fixtures, the energy supply must be modulated through a ballast. Taken together, the lamp, fixture, and any necessary ballast are sometimes called a *luminaire*.

Basic Concepts of Electricity

Electricity generally comes in two basic forms:

1. *Direct current (dc)*. Current that flows in a continuous direction (e.g., from a battery or directly from a solar panel or dc generator).
2. *Alternating current (ac)*. Current that rapidly changes the direction of voltage (e.g., power supplied by the grid). The direction can change many times per second, and the frequency of this change is measured in cycles per second, or hertz.

Although it hasn't been without controversy, utilities currently deliver electricity to us in ac form because it has less of a voltage drop across distances. The electricity we use then is measured by an electric meter. For many of us, these meters will soon be upgraded to *smart meters* in the next few years, which will provide real-time monitoring capabilities and additional efficiencies.

Let's take a closer look at electricity:

- *Voltage (V)*, measured in volts, is commonly understood as the pressure pushing the current.

- *Resistance (R)*, measured in ohms, is a material's resistance to electrical flow.
- *Current (I)*, measured in amperes, is the amount of electricity. $I = V/R$; that is, current is equal to voltage divided by resistance.

An analogy commonly used to explain electricity, which is invisible, likens the phenomenon to plumbing. There are some limitations to this exercise, but it can provide some understanding:

- Voltage (*V*) is represented as water pressure.
- Resistance (*R*) is represented as pipe size.
- Current (*I*) is represented as the flow rate.

Another important concept when it comes to electricity is power.

- Electrical power (*P*) is the rate at which electrical energy is transferred by an electric circuit. This is measured in watts. In simple circuits, an electrical power system *P* is equal to voltage times current: $P = VI$.
- Watts = volts × amps
- 1 Kilowatt = 1,000 watts

Utility Billing

Electricity use typically is measured in units called *kilowatt-hours* (kWh). One kilowatt-hour is equal to 1,000 watts of power used over a period of one hour—for example, ten 100-watt light bulbs turned on for one hour. Your monthly electric bill is calculated by multiplying the cost of one kilowatt-hour, set by your utility, by the number of hours of electricity you used. The price per kilowatt-hour usually changes according to the season and time of day, as well as market prices, although many consumers are charged flat rates, with the utility bearing the brunt of the minute-by-minute fluctuations.

Beware: You can't directly convert kilowatts into kilowatt-hours because that would be like converting miles per hour into miles traveled; you need to know how long you traveled at that speed.

Lighting Basics

- *Lumen (lm)*. The standardized SI (International System) unit of measurement for luminous power, which is the perceived power of a light source. Luminous power is also called *luminous flux*, and it is adjusted to reflect the varying sensitivity of the human eye to different wavelengths of light. Technically, one lumen is defined as the luminous flux of light produced by a source that emits one candela of luminous intensity over a solid angle of one steradian. Don't worry if you don't fully understand this, though; what's important for our purposes is that more lumens means stronger light.

– *Initial or rated lumens.* The light output of a lamp for the first 100 hours of use.

– *Mean or design lumens.* Estimated light output of a lamp between 100 hours of use and 75 percent of lamp life.

• *Lumens per watt (lm/W or LPW).* A measure of luminous *efficacy*, which is essentially efficiency. This is the ratio between the light output (called luminous flux) emitted by a device and the amount of input power it consumes. It is also sometimes referred to as the *wall-plug luminous efficacy* or *wall-plug efficacy*. A typical candle has an overall luminous efficacy, in lumens per watt, of 0.3. Standard incandescents range between 13 and 17 lumens per watt, and halogens are a bit higher, up to around 24 lumens per watt. LEDs normally range from 50 lumens per watt to the low 90s (higher in some recent examples), and fluorescents range from the middle 40s up to 100 lumens per watt. Low-pressure sodium lamps, often used for streetlights, range from 100 to 200 lumens per watt, but they don't render colors well, as we'll see.

• *Candela (candlepower, abbreviated as cd).* A unit of measurement of light intensity at various angles. The candela is the SI base unit for luminous intensity; a candle emits light of roughly one candela.

• *Foot-candle.* Lumens per square foot striking a surface. Basically, this is the amount of light that a single candle would provide to a one-foot-radius sphere.

• *Intensity.* A measure of the time-averaged amount of light striking a given area. For bulbs, this is measured in lumens, and for fixtures, it is measured in *lux* (lumens per square meter). The higher the lux, the more light a fixture produces.

• *Lumen maintenance.* How well a light source retains its intensity over time. HID lamps have poor lumen maintenance, losing up to 40 percent of their intensity within six months. A quality fluorescent bulb loses only 10 percent of its intensity over that time.

- *Rated life(span)*. A measure of the amount of time a bulb is expected to last. Technically, the rated life of incandescents and fluorescents is determined by measuring how long it takes for half the products in a test group to fail. This standard doesn't apply to LEDs, however, because they tend to fail gradually. The industry has decided that the "burn out" of an LED occurs when it has dropped to 70 percent of its initial output. It's also true that many LEDs haven't been on the market long enough to really make an accurate prediction of lifespan; in this case, reported figures are often extrapolated.
- *Visual acuity*. The eye's ability to detect fine details.
- *Fixture efficiency*. The ratio of light emitted from a fixture versus the light emitted by just the lamp(s) in the fixture, expressed as a percentage.

Color Temperature and Color Rendering Index

An important principle of lighting is *color temperature*, which has a direct effect on our mood and what we perceive as the quality of light. In the Green Depot flagship store on the Bowery in New York City (in the historic lighting district!), shoppers can try out the different color temperatures of bulbs in a test booth. There's quite a range of light qualities, from regular incandescents to different CFLs, halogens, LEDs, and cold-cathode CFLs.

- *Correlated color temperature*. The appearance of the color of light to the human eye, expressed in Kelvins (K), the SI unit for temperature. The higher the color temperature, the cooler the light will be, and the lower the color temperature, the warmer the source will appear to be.
  - *Cool light* is typically preferred for visual tasks, such as those completed at a workbench or desk, on a factory floor, and so on.
  - *Warm light* is often preferred for living spaces and provides better skin tones.

As shown in Table 2-1, the colors of light range from 1,700 to 9,300 K. Light at 5,000 K, often called *brilliant white*, is the approximate color temperature of daylight at about 11:00 a.m. in New York City, whereas light at 6,500 K would be daylight at approximately

TABLE 2-1 Color Table for Light *(DOE/Wikipedia)*

Temperature	Source
1700 K	Match flame
1850 K	Candle flame
2700–3300 K	Incandescent light bulb
3350 K	Studio "CP" light
3400 K	Studio lamps, photofloods, etc.
4100 K	Moonlight, xenon arc lamp
5000 K	Horizon daylight
5500–6000 K	Typical daylight, electronic flash
6500 K	Daylight, overcast
9300 K	CRT screen

Note: These temperatures are merely approximations; considerable variation may be present.

2:00 p.m. Color temperature also can be given in mireds, which are primarily used by lab scientists. A *mired,* short for "microreciprocal degree," is defined as one million divided by the color temperature in Kelvins. For those who might not remember from high school or college chemistry, the Kelvin temperature scale has the same interval as Celsius but has the zero point at *absolute zero,* where all molecules stop moving.

0 degrees Celsius = 273.15 Kelvin = 32 degrees Fahrenheit

As shown in Table 2-2, color temperature ranges are often associated with specific lighting names (*soft white, bright white,* etc.). However, these colors are not standardized for modern bulbs like

TABLE 2-2 Color Chart for Lighting *(DOE/Wikipedia)*

Color temperature	Kelvin	Mired
'Warm white' or 'Soft white'	≤ 3000 K	≥ 333 M
'White' or 'Bright white'	3500 K	286 M
'Cool white'	4000 K	250 M
'Brilliant white or Daylight'	≥ 5000 K	≤ 200 M
Daylight	6500 K	

they were for older-style (halophosphate) fluorescent lamps. This can make it more difficult to compare lighting quality across brands, unless color temperatures are referenced, because variations and inconsistencies exist. It's always best to try out a bulb before you buy or try one and see how it works before you buy several.

For example, Osram Sylvania's Daylight CFLs have a color temperature of 3,500 K, whereas most other lamps with a "daylight" label have color temperatures of at least 5,000 K. Some vendors do not include the Kelvin value on the package, but this is beginning to change now that the Energy Star standard for CFLs is expected to require such labeling.

- *Color rendering index (CRI).* This measures a light source's ability to render colors accurately. The CRI scale goes from 1 (the lowest) to 100 (able to render colors completely, as with sunlight). A CRI of 75 is often considered good for lighting, with 85 being very good and 95 excellent.

  Some manufacturers label their CFLs with a three-digit code to identify the CRI and color temperature. In this case, the first digit represents the CRI measured in tens of percent, whereas the second two digits represent the color temperature measured in hundreds of Kelvins. For example, a CFL with a CRI of 83 percent and a color temperature of 2,700 K would be given a code of 827.

Daylighting

Daylighting refers to the use of windows and skylights to bring sunlight into a building. It is a twentieth-century term applied to design elements that have been used for millennia but which were deemphasized in an age of relatively cheap fossil fuels and lack of concern for the environment. The good news is that today's highly energy-efficient windows, plus advances in lighting design, have helped to make it possible to reduce the need for "artificial lighting" during daylight hours. And done well, daylighting need not overly interfere with heating or cooling the house—cool! More on this in Chapter 8.

Green Building Labels

When it comes to building with green, environmentally friendly features, there can be a lot of noise and confusion about what's really sustainable. As a result, various green building certification programs have arisen around the world. Many are locally focused, which offers the benefit of maximizing for local climactic and cultural variables. But this also can be confusing to consumers, especially because people move around frequently. Here's a look at some of the most important green building programs.

NAHB National Green Building Program

The National Association of Home Builders (NAHB) offers a National Green Building Program in which new homes can be awarded bronze, silver, or gold levels (Figure 2-2). Each home gets scored on a checklist of seven categories: lot design, resource efficiency, energy efficiency, water efficiency, indoor environmental quality, operation and maintenance, and global impact. Scores are verified by an NAHB-accredited third party.

Some observers have complained that the NAHB program is too easy for builders to meet. The centerpiece New American Home 2008 was certified to gold level, even though it was a colossal 7,000 square feet. The home is 62 percent more energy efficient than conventional models of the same size in the same climate, although some critics argue that something so large shouldn't ever be labeled green.

FIGURE 2-2   The National Association of Home Builders (NAHB) offers a National Green Building Program.

EPA's Energy Star for Homes

The Environmental Protection Agency's (EPA's) blue-and-white Energy Star logo (with accompanying yellow label) has become quite well known on appliances, electronics, and even lighting, and has been approved for more than 18,000 products. Not many people know that they also can shop for Energy Star–qualified homes, though. Energy Star for Homes focuses exclusively on energy efficiency, and the guideline is that a participating structure must be at least 15 percent more efficient than homes built to the 2004 International Residential Code (IRC). In reality, many of the homes that so far have been certified are quite a bit more efficient (Figure 2-3).

FIGURE 2-3   The EPA's Energy Star program is widely known on consumer products, but less so on whole homes.

USGBC's LEED

The U.S. Green Building Council's (USGBC's) Leadership in Energy and Environmental Design (LEED) standards have won high praise for rigorous green guidelines, first in the commercial sector and then later with homes. LEED takes a holistic approach that encompasses energy efficiency, indoor air quality, material use, water conservation, landscaping, and more. For homes, large sizes incur penalties. Buildings can earn the coveted levels of certified, bronze, silver, gold, or platinum (Figure 2-4).

FIGURE 2-4 Leadership in Energy and Environmental Design (LEED) standards have won high praise for rigorous green guidelines.

Specific categories include LEED for New Construction, LEED for Existing Buildings, LEED for Commercial Interiors, LEED for Retail, LEED for Schools, and LEED for Core & Shell rating systems (as well as LEED for Homes). Many building types may qualify, from offices to stores, restaurants, hotels, libraries, schools, museums, apartment buildings, and religious institutions. According to the USGBC, LEED-qualified homes cost only 2 to 5 percent more than conventional homes, but they result in substantial energy savings (at least 15 percent over comparable homes).

ISO 14000

A program of the International Organization for Standardization, ISO 14000 is intended as a tool for commercial facilities to assess and improve their environmental impact and to communicate their commitment to others. The program is really a process that guides managers through sustainability questions and decisions. As part of achieving ISO 14000 compliance, companies often will take steps to green up their lighting (Figure 2-5).

FIGURE 2-5 ISO 14000 certification from the International Organization for Standardization is intended as a tool to help commercial facilities maximize efficiency.

Summary

Although the physics and chemistry of lighting can be complicated and are not fully understood by scientists, there are only a few variables that most of us need to be concerned with when it comes to illuminating our lives. These include color temperature and rendering, wattage, light output, lifespan, and a few others. As we will soon see, much of lighting is intuitive, and it is easy to experiment along the way, as well as tap into our creativity.

In general, our goals always should be to produce the light we need as efficiently as possible by choosing lower wattages, longer lifespans, and better fixtures. This saves us money as well as environmental impact.

CHAPTER 3

Incandescent, Halogen, and Gas-Discharge Lighting

When most people think of a light bulb, they envision the classic *A-shaped* incandescent bulb, although this is gradually starting to change with the proliferation of compact fluorescent light bulbs (CFLs) and other more efficient technology, not to mention impending legal restrictions. Still, the basic incandescent bulb hasn't changed much in the 120 years since Thomas Edison invented a profitable model (Edison did not invent the incandescent bulb, he just made one that sold well). Today, incandescent lighting is still the most common type of illumination used in homes, where it traditionally has delivered about 85 percent of household lighting. In fact, according to *Fast Company*, 425 million incandescent bulbs are sold in the U.S. every year, making up roughly half of the bulb market. As you will see, however, *this is the least green bulb out there*.

Incandescent bulbs light up instantly, providing warm light and excellent color rendition. They dim easily, which reduces energy use as well as light output. Incandescents work well with alternating current (ac) and direct current (dc), and they are inexpensive. Most existing light fixtures are designed for the size and shape of these traditional bulbs, and everyone is familiar with them, from interior decorators to contractors and home owners.

History and Technological Overview

Incandescent bulbs create light by running an electric current through a resistive filament, which today is commonly a tiny coil of tungsten. The filament gets so hot that it glows and produces visible light—although 90 to 98 percent of the energy released comes in the form of wasted heat. These bulbs are named after the principle of *incandescence*, which is a term for the process of emitting light on heating to high temperatures (Figure 3-1). In contrast, fluorescent lights, high-intensity-discharge (HID) lamps, and light-emitting diodes (LEDs) heat by *luminescence*, which occurs at lower temperatures.

Although Thomas Edison is widely credited with inventing the incandescent light bulb, scientists had been working on the technology long before the famous inventor was born. In fact, a recent scholarly review of the technology's history named at least 22 people who contributed to the invention. In the first years of the 1800s, Sir Humphry Davy produced incandescent light at the Royal Institution of Great Britain. Davy hooked up a powerful battery to a thin strip of platinum, which glowed dimly. A philosopher as well as the leading chemist of his day, Davy discovered a number of important chemical elements through electrolysis, including sodium, potassium, calcium,

FIGURE 3-1   Incandescents have had a long run, but their inefficiency will be their downfall. The heat under these Las Vegas lights is stifling. *(Photo by Brian Clark Howard)*

magnesium, and barium. He also discovered the useful medical properties of nitrous oxide, commonly called "laughing gas."

Davy demonstrated the first arc light, either in 1807, 1808, or 1809, depending on the source. He used two charcoal sticks as electrodes, and when he connected his 2,000-cell battery, a bright white light arced across the four-inch gap between the sticks. Davy originally dubbed the effect an "arch light" owing to its shape, but the name was soon shortened to "arc light." Carbon arc lights were produced and sold, although they emitted uneven light and produced a great amount of heat. Today, arc lights are still used in some applications in medicine and movies.

Davy also developed the miner's safety light (often called a *Davy lamp*), which could be used around natural gas without causing explosions owing to a heat-dissipating baffle. Unfortunately, Davy suffered a number of accidents and exposures to toxic chemicals in his work, and he died relatively young, at age 51, most likely as a result. However, his assistant—Michael Faraday—continued the work of expanding our understanding of electricity and became famous in his own right.

A number of others worked on making lighting brighter and more reliable over the years. In 1878, British scientist Sir Joseph Wilson Swan was granted a patent for an incandescent light bulb, and his house in England became the first in the world to have electric lighting. Swan used a carbonized thread in a bulb with a partial vacuum so that the filament wouldn't oxidize. (Years later, engineers discovered that the light quality improved if the bulb was filled with an inert gas such as argon or nitrogen instead of a vacuum, and this is the common practice today.)

Edison's first patent on incandescent lighting appeared a year after Swan's. Edison also designed bulbs with vacuums and tested thousands of different materials for filaments, eventually settling on carbonized cotton (Figure 3-2). Later, he would also use carbonized bamboo, which provided longer life. When Edison began mass producing bulbs in the late 1800s, Swan sued him for patent infringement. Swan won in court, and as part of the judgment, Edison had to take his rival on as a business partner in Great Britain.

Therefore, if Edison didn't really "invent the light bulb," why is he so famous? For one thing, he also invented a lot of other useful things, such as the phonograph and motion picture camera, ulti-

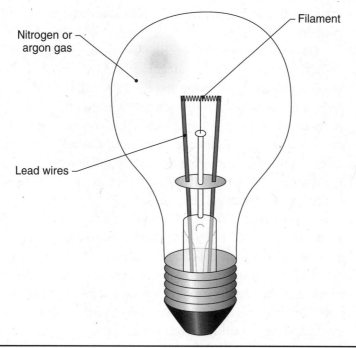

Nitrogen or argon gas

Filament

Lead wires

Figure 3-2   Inside a typical incandescent bulb. *(U.S. Department of Energy)*

mately holding more than 1,000 patents. He also did much to pioneer the modern industrial laboratory and was a successful businessman. Edison's version of the light bulb went mainstream owing to his business savvy as well as his quality product. In short, he produced a better light bulb than Swan or other rivals, and he presented it to the public in an integrated system of electric lighting that caught on rapidly and spread around the globe.

Problems with Incandescent Bulbs

Despite their colorful history, there are a number of drawbacks to incandescents. They have a short average operating life, 750 to 2,500 hours, with 1,000 hours being typical. This is part of the reason why they are often not selected for commercial applications because checking and changing light bulbs in large facilities adds significantly to labor costs. Incandescents are also inefficient because up to 98 percent of the energy used is emitted as waste heat. This excess heat also adds to cooling loads during warm months and can result in fire danger.

A 100-watt, 120-Vac light bulb produces about 1,700 lumens, or about 17 lumens per watt. This is a low luminous efficacy compared with other lighting options. In comparison, LED lamps range from 50 lumens per watt to the low 90s, and fluorescents range from the middle 40s up to 100 lumens per watt.

Types of Incandescent Bulbs

There are three common types of incandescent bulbs (Table 3-1):

1. Standard incandescent bulbs
2. Halogen bulbs
3. Reflector bulbs

TABLE 3-1   Typical Specifications of Different Incandescent Bulbs
(U.S. Department of Energy)

Incandescent Lighting Type	Efficacy (lumens/ watt)	Lifetime (hours)	Color Rendition Index (CRI)	Color Temperature (K)	Indoors/ Outdoors
Standard "A" bulb	10–17	750– 2500	98–100 (excellent)	2700–2800 (warm)	Indoors/ outdoors
Tungsten halogen	12–22	2000– 4000	98–100 (excellent)	2900–3200 (warm to neutral)	Indoors/ outdoors
Reflector	12–19	2000– 3000	98–100 (excellent)	2800 (warm)	Indoors/ outdoors

Standard Incandescent Bulbs (and Variations)

Known as the screw-in, A-type light bulb, standard incandescent bulbs are the most common light source available (Figure 3-3). They produce light by heating a tiny tungsten coil. The larger the wattage of an incandescent, the higher is the efficacy, although you'll use more power. In lighting, efficacy is expressed as the ratio of the amount of light produced (in lumens) by the amount of power consumed (in watts). This is technically not exactly the same thing as efficiency, which is always a dimensionless ratio of output divided by input—in the case of lighting this means watts of visible power divided by power consumed in watts. However, there isn't an easy way to measure watts of visible power, and estimating it is a tech-

FIGURE 3-3   Vintage-style, exposed-filament incandescent bulbs are trendy for restaurants and bars, but they use roughly three times the energy of standard incandescents. *(Photo by Brian Clark Howard)*

nical exercise, so for most practical applications it's useful to think of efficacy as essentially efficiency.

Note that so-called long-life incandescent bulbs come with a price, in addition to the higher purchase point. Built with thicker filaments, these bulbs are even less energy efficient and so are not favored by green designers or environmentalists.

Better are so-called EnergyMiser bulbs, sometimes called *supersaver* bulbs. These are incandescents designed to be more efficient, generally using 5 to 13 percent less electricity, with only minimal decreases in light output. They tend to cost a bit more than standard bulbs as well, but they may pay for themselves over time. These bulbs may be good choices in fixtures that can't readily accept a more advanced product, such as a CFL.

In recent years, in no small part because of federal lighting efficiency standards, there has been fresh interest in designing more efficient incandescents. General Electric has announced ambitious plans in this area, with initial stated production goals of doubling incandescent efficiency in the short term. Time will tell if these developments are successful and price competitive.

Halogens

Well known today, halogen bulbs are a specialized type of incandescent lighting, and they achieve better energy efficiency. In a halogen lamp, a tungsten filament is sealed inside a small transparent envelope that is then filled with an inert gas plus a small amount of a halogen, which is an element from the column of the periodic table that includes iodine and bromine (remember high school chemistry?). The halogen and filament undergo a chemical reaction called a *halogen cycle* that redeposits tungsten back onto the filament (Figure 3-4).

In normal incandescents, tungsten gradually settles on the inside of the bulb, and the filament becomes thinner and more brittle over time. Eventually, it wears too thin and breaks, which is the most common way bulbs fail. In halogens, however, the redepositing of tungsten prolongs the life of the filament. The halogen cycle also helps to prevent the bulb from darkening.

In a halogen lamp, the filament heats to a hotter temperature than in standard incandescents, and this gives it a higher efficacy (10 to 30 lumens per watt). Overall, halogens are 10 to 40 percent more efficient than incandescents, and they tend to last two to three times longer. The light provides excellent color rendition, and it has a higher color temperature than incandescents.

FIGURE 3-4   A lit halogen bulb. Notice the glowing filament. *(Photo by Brian Clark Howard)*

However, halogens are more expensive than standard incandescents. Since many aren't that much more efficient, they may not result in very substantial savings over time. The California Energy Commission calculated that using a single 300-watt halogen lamp for 2 hours a day would consume 220 kilowatt-hours of electricity in a year at an average annual cost of about $18. On the plus side, halogens can be made in small sizes but still shine brightly. This has made them popular for auto headlights and in track lighting, in which one can use a lower-wattage bulb but get an attractive effect.

However, halogens have a number of drawbacks. The outer bulb is made of quartz or special glass, but it still can shatter violently, posing a risk of cuts. And halogens can produce a lot of heat, adding to cooling loads. This is particularly true for torchiere-style floor lamps that use halogen bulbs rated at 300 watts or more. As a kid, one of us had an 800-watt halogen torchiere, but it heated up the bedroom so uncomfortably that it was rarely switched on during the warm months. A 500-watt halogen lamp produces four times more heat than a typical incandescent and reaches temperatures of 1,200°F. That kind of heat can ignite fabrics on contact, which is why halogens are often blamed for starting fires and are banned from many dorm rooms.

Quartz halogen lamps also can release a substantial amount of ultraviolet (UV) radiation, enough so that unshielded bulbs even could produce a sunburn. This is rarely a concern in practice, however, because these are normally protected with doping compounds or UV shield glass to block the harmful rays.

It's also important not to touch the quartz envelope of a halogen bulb. Any surface contamination, especially fingerprints, can create a hot spot when the bulb is heated. This can lead to weakness, which can result in dangerous shattering. As a precaution, manufacturers suggest handling the bulbs with a clean paper towel or only touching on porcelain parts. Any fingerprints should be carefully removed with rubbing alcohol, which must be dried off before the bulb is used.

Many writers exclude halogens from lists of green lighting because they produce so much excess heat and because they aren't that much more efficient than conventional incandescents. Still, halogens can have their place in a well-designed lighting scheme, and in some cases they are preferable to standard bulbs. In many ways, they can be a transitional technology to be used over the next few years before LEDs and other more advanced replacements become more available and affordable.

Perhaps symbolically, from 1999 to 2006, halogen bulbs were used on the iconic Times Square ball that marks the New Year in the Big Apple. But starting in 2007, however, they were replaced with LEDs.

Halogen Energy Savers

A newer class of halogens being marketed to an energy-conscious public is so-called halogen energy savers, which bring the greater efficiency of halogen lighting to standard incandescent sizes and shapes. These bulbs typically are 30 percent more energy efficient than regular incandescents, but they still produce considerable excess heat and can be a fire hazard.

One example is the halogen SuperSaver by Osram Sylvania, which uses only 43 watts to replace a 60-watt incandescent (Figure 3-5). Another leading brand is the Philips Halogena Energy Saver, which uses a special chamber to reflect wasted heat back to the filament, boosting lighting power (Figure 3-6).

FIGURE 3-5   Sylvania SuperSaver. *(SMSB Consulting Group, Inc.)*

FIGURE 3-6   Philips Halogena Energy Saver. *(1000bulbs.com)*

These bulbs have the following features:

- 22 to 47 percent energy savings over standard incandescents
- Contain no mercury
- Bright, white light
- Last at least two years
- Fully dimmable
- Instant on
- Come in A-shape, decorative, and flood

Other Types of Halogens

These days, halogens come in many shapes and sizes. Some of the most common include:

- Automotive and vessel
- Commercial applications
- Projection lamps
- Linear "double-ended" lamps

- Track lighting
- Replacement bulbs for standard fixtures
- Reflector lamps

Reflector bulbs direct and spread light over specific areas, as opposed to in all directions, like conventional bulbs. Reflectors are used mainly for floodlighting, spotlighting, and downlighting. They come in two basic types: parabolic aluminized and ellipsoidal.

Parabolic aluminized reflector bulbs (type PAR) typically are used for outdoor floodlighting. Ellipsoidal reflector bulbs (type ER) focus light about two inches in front of the enclosure, which makes them popular for recessed fixtures. Ellipsoidal reflectors are twice as energy efficient as parabolic reflectors for recessed fixtures.

Low-Pressure Sodium Lighting

Ever notice that many streetlights have a yellowish tint, or that people look strange and sickly underneath them? You're seeing the effects of *low-pressure sodium* lights. The good news is that this technology is extremely energy efficient, with luminous efficacy of 100 or even 200 lumens per watt. The lights are so efficient in part because they emit a wavelength that is near the peak sensitivity of the eye. However, they are extremely poor at color rendition—hence the sickly pall effect.

Low-pressure sodium lights last a long time—18,000 hours is common these days—and are used frequently in highway and security lighting, where color isn't important. Like high-intensity discharge lighting, low-pressure sodium bulbs require up to 10 minutes to start up, and they have to cool before they can restart. Therefore, they are really suitable only where they will stay on for hours at a time, and they obviously won't work with motion detectors.

The core of these lamps is a borosilicate glass gas-discharge tube that is filled with solid sodium and a small amount of neon and argon gases. The light is produced when the sodium is heated and vaporizes.

Low-pressure sodium lamps are rarely used in residential settings, although they have been workhorses in public areas, where they can be part of an efficient lighting strategy.

High-Intensity-Discharge Lighting

Like the arc light demonstrated by Humphry Davy in the early 1800s, modern *high-intensity-discharge* lamps (HIDs) create light by forming an arc of electricity between two electrodes. Only in the case of HIDs, the electrodes are sealed in a tube that is filled with a mercury, sodium, or metal halide gas as the conductor, as shown in Figure 3-7.

HID bulbs provide long life and high efficiency, with luminous efficacies from 65 to 150 lumens per watt. They can reduce lighting energy use by 75 to 90 percent when they replace incandescent bulbs, according to the Department of Energy (DOE). Of course, the application must be right.

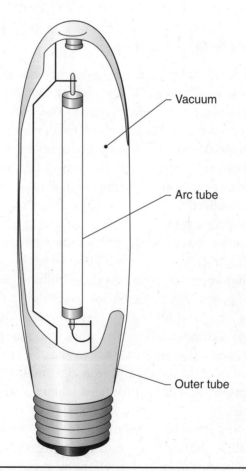

FIGURE 3-7   HID lighting. *(U.S. Department of Energy)*

Like fluorescent bulbs, HIDs require ballasts. And they take up to 10 minutes to produce light when first switched on because the ballast needs time to establish the electric arc. HIDs produce bright light and are used commonly for outdoor lighting and in large indoor arenas. These are often the bright lights that shine down on baseball and soccer fields. Since the bulbs take a while to establish, they are most suitable for applications in which they stay on for hours at a time (Figure 3-8).

These are the three most common types of HID bulbs:

1. Mercury vapor bulbs
2. High-pressure sodium bulbs
3. Metal halide bulbs

Mercury Vapor Bulbs

Mercury vapor bulbs—the oldest type of HID lighting—are used primarily for street lighting now. They provide 26 to 60 lumens per watt and cast a very cool blue/green-white light. These days, mercury vapor bulbs are used less commonly in arenas and gymnasiums.

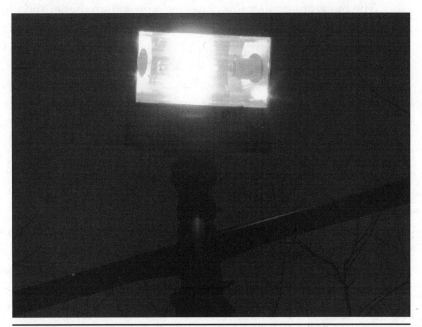

Figure 3-8   Discharge lighting is used most commonly outdoors in public areas. *(Photo by Brian Clark Howard)*

Metal halides are often chosen instead because they have better color rendering (a color rendering index of 70 versus 50 for mercury vapor), as well as higher efficacy (70 to 115 lumens per watt). However, mercury vapor bulbs have longer lifetimes (16,000 to 24,000 hours) than metal halide bulbs (Table 3-2).

TABLE 3-2 A Comparison of HID Lighting (*U.S. Department of Energy*)

High-Intensity Discharge Lighting Type	Efficacy (lumens/ watt)	Lifetime (hours)	Color Rendition Index (CRI)	Color Temperature (K)	Indoors/ Outdoors
Mercury vapor	25–60	16,000– 24,000	50 (poor to fair)	3200–7000 (warm to cold)	Outdoors
Metal halide	70–115	5000– 20,000	70 (fair)	3700 (cold)	Indoors/ outdoors
High-pressure sodium	50–140	16,000– 24,000	25 (poor)	2100 (warm)	Outdoors

High-Pressure Sodium Bulbs

High-pressure sodium lighting is becoming the most common type of outdoor lighting. It is also used commonly to grow plants indoors. The bulbs are smaller than low-pressure sodium lights and also include mercury. The arc tube is often made of translucent aluminium oxide. High-pressure sodium lighting has an efficacy of 50 to 140 lumens per watt—an efficiency exceeded only by low-pressure sodium bulbs. Unlike low-pressure sodium bulbs, though, the high-pressure bulbs produce a warm white light, around 2,100 K. Like mercury vapor bulbs, high-pressure sodium bulbs have poorer color rendition (CRI of 25) than metal halide bulbs but longer lifetimes (16,000 to 24,000 hours).

Metal Halide Bulbs

Metal halide bulbs contain mercury vapor as well as metal halide vapor, which allows them to produce more light with more lumens per watt (Figure 3-9). Metal halides have a bright white light in the cold spectrum (3,700 K). They also have the best color rendition among HID lighting types, with a CRI around 70. They are used in large indoor areas, such as in gymnasiums and sports arenas, as well as outdoors, such as in car lots.

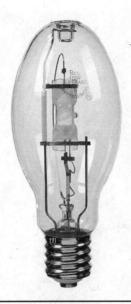

FIGURE 3-9   A metal halide bulb. *(Lithonia Lighting)*

Metal halide bulbs have shorter lifespans (5,000 to 20,000 hours) than either mercury vapor or high-pressure sodium bulbs, according to the DOE (Figure 3-10).

Summary

We have learned that incandescent lighting has a number of important limitations, notably short lifespans and poor energy efficiency—although researchers are trying to make improvements. Halogen technology is an improvement on incandescents, yet halogens have their own drawbacks, including high heat release. Low-pressure sodium and HID lighting offer long life and good efficiency but poor color rendering and long startup times.

The lighting all around us offers many opportunities to do things differently.

FIGURE 3-10   HID lighting is responsible for the "bright lights" of sports venues. *(Photo by Brian Clark Howard)*

CHAPTER 4

Fluorescent Lighting

The distinctive swirl of the compact fluorescent light bulb (CFL) arguably has become one of the most pervasive and successful symbols of the green movement (Figure 4-1). Thousands of CFLs have been passed out for free by activists and corporations, and the bulbs have been subsidized by governments and utilities. As countries and organizations have taken a stand against less efficient incandescent light bulbs, CFLs have been widely seen as their natural successors. In fact, CFLs have been promoted and hyped so much that they have also become a bit of a cliché, and leaders in the environmental movement are now spending less air time talking about them in favor of other initiatives that appear to "go beyond just changing a light bulb."

In some ways, this is unfortunate, because adoption of CFLs is slowing down in the United States despite the tougher lighting energy standards that are gradually coming into effect. So far, the high-water mark was 2007, when the U.S. Environmental Protection Agency (EPA) reported that 290 million CFLs were sold that year, nearly double the total for 2006. In 2007, CFLs accounted for about 20 percent of the U.S. light bulb market. This figure was aided by the fact that *An Inconvenient Truth* came out in 2006 and won an Oscar, and Hurricane Katrina had devastated the Gulf Coast in late 2005, an event that was being widely linked to climate change. Many people were pointing to CFLs as an easy solution, and the bulbs received major promotion from retailers, including Walmart, Home Depot, Costco, and others.

FIGURE 4-1   A now-ubiquitous spiral CFL. *(Photo by Brian Clark Howard)*

Since then, sales of CFLs have declined by 25 percent, and the U.S. Department of Energy (DOE) recently estimated that 90 percent of eligible fixtures in the country still sport incandescent bulbs. Why? There are likely a number of reasons, not the least of which is the global recession. Even though CFLs save consumers money in the long run (the average payoff period is estimated at two years), they still cost a bit more than standard incandescents up front, and cash-strapped people are more likely to make buying decisions based on initial sticker price.

It's perhaps not surprising that the bulbs also were overhyped, and shelves were flooded with cheap versions hastily assembled in China (where about 90 percent of light bulbs are made). Many consumers found that the CFLs they bought at drug and dollar stores didn't last nearly as long as boosters promised, and many complained that they didn't like the "harsh" light or the nontraditional shape. Scare campaigns also flooded the Internet, exaggerating the mercury content of the bulbs (more on this later) and implicating them as a cause of migraines and seizures, which is dubious scientifically. Some impatient consumers reverted back to their old ways, particularly as climate change deniers gained new ground in the wake of overblown "scandals." (Of course, it is also possible that the longer life of CFLs has meant consumers have had to shop for replacement bulbs less often.)

To be clear, CFLs are one of the best choices for green lighting that we have today. Good-quality CFLs now cost around $2 to $4, which is a substantial drop from the $15 to $20 we remember from just a few years ago. CFLs use about 75 percent less energy than incandescent bulbs to produce the same amount of illumination, for a luminous efficacy of 30 to 110 lumens per watt. This is largely possible because fluorescents produce two-thirds to three-quarters less waste heat and convert about 22 percent of input power to visible light. CFLs should last 8 to 15 times longer than incandescents (10 times is commonly quoted). Improvements in technology have resulted in fluorescents with color temperature and color rendition that are now comparable with those of incandescent bulbs.

To guarantee quality, it's a good idea to look for Energy Star–registered CFLs, which must meet minimum standards for lifespan, brightness, efficiency, and low mercury content. It's also smart to buy major brands, such as Philips, Sylvania, Litetronics, and General Electric (GE). According to the EPA, a typical CFL can save the user over $30 in energy costs over the course of the product's life. Note that some writers list the savings at $47 per bulb, the Environmental Working Group recently estimated the savings at up to $80, and New York City advertises that each bulb can save $100—of course, it all depends on electric rates and usage patterns.

Technological Overview

Fluorescent lights as we know them were invented in the 1930s, and the essential technology hasn't changed that much since. Fluorescents are technically gas-discharge lamps with the added feature of a phosphor coating. When you turn on a fluorescent light, electrons "boil off" the coating material of the electrodes at the ends of the tubes. An electric arc forms between the electrodes, and ionizes the mercury vapor in the tube, which is kept at low pressure, typically less than one percent of atmospheric pressure. Mercury atoms are excited, and when they drop down to a lower, more stable energy state, they emit short-wave ultraviolet (UV) light. This strikes the phosphor coating on the inside of the tube, causing it to fluoresce and produce visible light. Eventually, the coating on the electrodes gets depleted, and the light fails, often turning black at the ends.

A fluorescent light's cathode (the negative electrode, where electrons originate) typically is made of coiled tungsten coated with a mixture of barium, strontium, and calcium oxides. The phosphor coatings on the tube regulate the color and quality of the light emitted and are made with varying blends of metallic and rare-earth phosphor salts (Figure 4-2). The coatings, along with the glass of the tube, prevent the UV light from escaping, where it could be harmful. In fact, a 1993 study concluded that UV exposure from sitting under fluorescent lights for eight hours is equivalent to only one minute of sun exposure.

Fluorescent bulbs also need a ballast to regulate current and provide startup voltage. Newer electronic ballasts are more energy efficient than previous magnetic ballasts, and they operate at a very high frequency, eliminating the flicker and noise that in the past gave

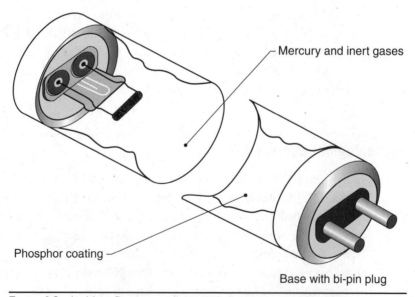

Mercury and inert gases

Phosphor coating

Base with bi-pin plug

FIGURE 4-2 Inside a fluorescent light. *(U.S. Department of Energy)*

fluorescents a bad name. In addition, fluorescents now can be dimmed, for the most part, with the help of special ballasts.

History of Fluorescent Lighting

Fluorescent lighting is based on centuries of experimentation and scientific discovery. The principle of fluorescence, in which a material emits light of a different wavelength than it absorbed, had been observed in certain rocks for centuries and was named after the mineral fluorite in the 1840s by Irish scientist Sir George Stokes. Around that time, scientists also had discovered that an electric current passed through a partially evacuated glass vessel would produce a radiant glow. The emerging understanding of such phenomena built on the explorations of electricity by Michael Faraday and others.

In the 1850s, a German physicist and glassblower named Heinrich Geissler created a mercury vacuum pump that could evacuate a glass tube efficiently. When an electric current then was passed through the tube, a strong green glow would develop near the cathode end. So-called Geissler tubes were born and were produced for decades for research, demonstration, and entertainment purposes. Different colors could be produced with different combinations of rarified (thinned) gases, such as neon and argon, as well as ionizable materials, such as sodium and different metals. Increasingly elaborate tubes were created, some with amusing shapes. Geissler tubes were a forerunner to neon sign lighting as well as fluorescents.

In the 1890s, two giants of innovation, Thomas Edison and Nikola Tesla, worked on early versions of fluorescent lighting. Both contributed to understanding of the technology, although neither produced a product that attained commercial success. Edison tried a coating of calcium tungstate, excited to fluoresce by x-rays, but decided not to produce it. Tesla demonstrated his fluorescent lamps at the 1893 Columbian Exposition in Chicago, delighting visitors to the "White City." At the fair, Tesla also lit a pavilion with electric lights on alternating current (ac), a collaboration with George Westinghouse that introduced many to the new technology.

Also in the 1890s, a former Edison employee by the name of Daniel McFarlan Moore had some success with tube lights that used carbon dioxide or nitrogen. However, Moore's inventions were expensive, difficult to install, and required high voltages, so they did not earn wide adoption.

In 1901, a well-connected electrical engineer by the name of Peter Cooper Hewitt was granted a patent for mercury vapor lights. Cooper Hewitt was the son of a mayor of New York City and grandson of industrialist Peter Cooper, who is best remembered for building the first steam locomotive in the United States and for founding the Cooper Union school. Cooper Hewitt's lights weren't the first to use mercury vapor, and they were dim and emitted a blue-green light, but they were produced in standard sizes and found some success in photography and industrial applications.

In 1926, Jacques Risler received a French patent for applying fluorescent coatings to neon light tubes, which had been invented a number of years earlier. Although Edison had already experimented with fluorescent coatings, their popularization in neon advertising signs helped to set the stage for modern fluorescents. The next year, three German scientists patented a high-pressure, low-voltage vapor lamp. Although this lamp never saw production, the patent was acquired eventually by GE, and it helped to inform later designs.

In the 1930s, researchers in London announced promising efficiencies with experimental green fluorescent lamps. This jumpstarted GE to take a more serious look at the technology. Two research teams were established at the company's historic lighting labs at Nela Park in Cleveland, Ohio, one led by George Inman and the other by Philip J. Pritchard. The teams made impressive progress and earned a key patent in 1941. GE brought to market practical, affordable fluorescents, and the technology took off during the war years and through the postwar boom.

The CFL was invented by lighting engineer Edward E. Hammer in 1976. At the time, Hammer was an engineer with GE at Nela Park, and he had been working on ways to save energy as a result of the oil crisis in the early 1970s. At first, he improved the efficiency of tube fluorescents, and then he turned his attention to something that would mimic a bulb shape. Hammer ultimately was successful, but GE decided not to market his innovative design for a CFL. According to Hammer's current Web site, drop-the-hammer.com, GE didn't think it was worth investing the money into new manufacturing facilities for the product. However, Hammer's idea eventually leaked out, and other companies started producing CFLs in the

1980s. At first they appeared as a trickle, but they gradually gained steam through the 1990s and into the 2000s.

Today, Hammer is retired, but he still educates the public about efficient lighting through podcasts and interviews. In 2002 he was awarded the Edison Medal by the Institute for Electrical and Electronics Engineers. His prototype for the CFL is now in the Smithsonian.

Now let's take a closer look at different types of fluorescent bulbs.

Fluorescent Tube and Circline Bulbs

Traditional tube fluorescent fixtures are sometimes called *liner fluorescents*, although they also can come in U shapes. They are the workhorses in commercial lighting (Figure 4-3). Recently, Hearst Corporation in New York City changed all 14,000 fluorescent tubes in its Leadership in Energy and Environmental Design (LEED)-certified 46-story world headquarters. According to the company's

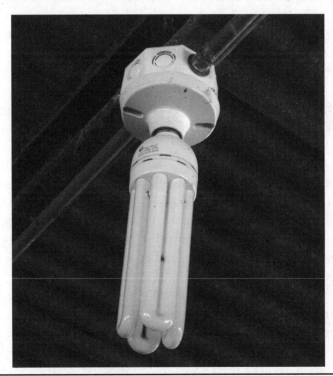

FIGURE 4-3   CFLs have found wide adoption around the world, including in this park shelter in rural Costa Rica. *(Photo by Brian Clark Howard)*

director of operations, Lou Nowikas, the upgrade will save $200,000 in energy costs a year and will pay for itself within just one year. The new bulbs use only 25 watts, compared with the 30 watts of previous bulbs, and are expected to last five years instead of the previous three.

Fluorescent bulbs are named by the diameter of the lamp. The major types include:

- *T5*. A fluorescent lamp that is 0.625 inch in diameter. These are among the newer models, and they have high energy efficiency and light output. They must work with an electronic ballast. They also tend to create more heat and have a shorter lifespan than other fluorescents.
- *T6*. A fluorescent lamp that is 0.75 inch in diameter. These are the newest types of fluorescent tubes, and they benefit from high energy efficiency and light output but produce less heat and last longer than T5s. They also require an electronic ballast.
- *T8*. A fluorescent lamp that is 1 inch in diameter. These are more energy efficient than older T12s, but they produce less light than newer T5s and T6s. They also require an electronic ballast to operate.
- *T10*. A fluorescent lamp that is 1.25 inches in diameter. These lamps aren't very common.
- *T12*. A fluorescent lamp that is 1.5 inches in diameter. These lamps were very common until recently; now they have been largely superseded by more efficient versions. They work with magnetic or electronic ballasts.

In order to determine the size of the tube in inches, simply divide the number of the name by 8 (12 divided by 8, in the case of a T12, is 1.5 inches). Note that T8, T10, and T12 fluorescent lamps typically are held with a bi-pin medium socket, whereas T5 and T6 lamps are usually held with a bi-pin minisocket. Also note that fluorescent tubes are sometimes further categorized by their lumen output as standard or high output (HO).

Fluorescent tubes are installed in a dedicated fixture with a built-in ballast. The two most common types are 40-watt, 4-foot (1.2-meter) bulbs and 75-watt, 8-foot (2.4-meter) bulbs. As you can see in Figure 4-2, in fluorescent tubes there is a very small amount of mercury and other inert gases to conduct the electric current. This allows the phosphor coating on the glass tube to emit light.

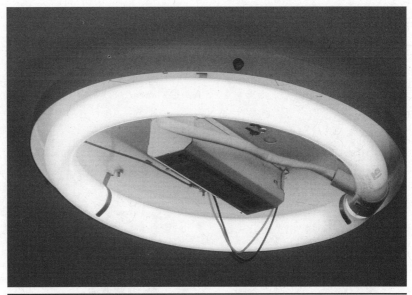

FIGURE 4-4   A circline fluorescent bulb in a kitchen. *(Photo by Brian Clark Howard)*

Tubular fluorescent fixtures and bulbs generally are preferred for ambient lighting in large indoor areas, such as schools, offices, and stores. In these situations, their low brightness creates less direct glare than incandescent bulbs.

Circular tube-type fluorescent bulbs are called *circline bulbs*. These are used commonly for portable task lighting, although they also can be used in overhead lighting as shown in Figure 4-4.

An important technical advance with fluorescents has been the replacement of older magnetic ballasts with electronic ballasts; this has removed most of the flickering and slow starting traditionally associated with fluorescent lighting. Magnetic ballasts are made of a magnetic core and coil and convert electric current to the proper voltage to run the lamp. In addition to being inefficient, magnetic ballasts produce a lot of heat; they also produce flicker owing to their 60-hertz operating cycle, which results in light flicker that's double this rate.

Electronic ballasts, in contrast, use circuitry to provide a stable current to the lamp regardless of input voltages. Note that this means that they usually won't work with dimmer switches unless the ballast is specifically designed to compensate. Inside an electronic ballast is a small circuit board, which usually includes recti-

FIGURE 4-5   A look inside a CFL electronic ballast. *(Anton/Wikimedia Commons)*

fiers, a filter capacitor, and two switching transistors (Figure 4-5). An electronic ballast applies a much higher frequency, around 40 kilohertz or more, which is why the lights don't flicker. The light is essentially continuous.

Still, it's important to note that the amount of efficacy lost through the ballast is significant, about 25 percent of lamp energy for magnetic ballasts and 10 percent of lamp energy for electronic ballasts. Also note that it's a good idea to use ballasts that carry the Underwriters Laboratories (UL) seal. Underwriters Laboratories writes testing standards, which are then checked at UL or one of a handful of other approved labs.

Another concept to be aware of is *ballast factor* (BF):

- *Low ballast factor.* This results in greater energy savings but reduced light.
- *High ballast factor.* This results in higher energy use but delivers more light.

Additionally, types of electronic ballasts include:

- *Rapid start.* Starts with a flicker or a blink before achieving full light.
- *Instant start.* Achieves full light as soon as the power is turned on.
- *Program start.* Ramps up power gradually to the bulb, although it may appear to give full light immediately. This is the best choice when there is frequent switching (on/off) or occupancy sensors are used.

The detailed electrical explanations for the different ways to start fluorescents are fairly complicated and are beyond the scope of this book.

Cold Weather Fluorescents

Fluorescents can be designed to work better in cold temperatures, typically with heavy glass jackets and/or krypton gas filling. These are sold for cold climates and for use in freezers, and they work best below room temperature. Note that all fluorescents should come with an operating temperature rating.

Fluorescent Tube Nomenclature

If you look at a typical T8 lamp, you will see something like this along the side of the bulb:

F 40 T12 / ES / RE-735

Let's break it down:

F = lamp type (FB/FU is for U-bent, and FT is for twin-tube T5).
40 = wattage (for slimline, high output [HO], very high output [VHO], and HOO lamps, this number corresponds to the length of the tube).
T12 = diameter of tube.
ES = modifiers (optional: ES is energy savings, HO is high output, VHO is very high output).
RE-735 = rare-earth phosphors. The first digit is the color rendering index (CRI), 70 in this example; the last two digits (optional) are color temperature (3,500 K in this example).

Compact Fluorescent Lighting (CFLs)

Compact fluorescent lighting (CFL) has been so successful in part because it is designed to replace regular incandescent bulbs and fit in existing light fixtures (Figure 4-6). In order to do this, a few design innovations were needed, including more powerful phosphors and folding of the tubes. Edward Hammer developed the spiral shape to try to pack in enough tube yet allow enough space between the coils to decrease blockage of light. Today, Hammer's spiral shape endures, although rectangular tubular-type CFLs are a bit more efficient and are more popular in Europe (Figure 4-7).

There are two main parts of a CFL:

1. The gas-filled tube, which is also called the *bulb* or *burner*.
2. The ballast, which emits and moderates electric current. This is usually part of the "bulb" but may be separate.

FIGURE 4-6   CFLs now come in a variety of shapes and sizes to fit almost any fixture. *(Photo by Brian Clark Howard)*

FIGURE 4-7 Tubular CFLs are slightly more efficient than spirals. They are more popular in Europe than in North America. *(Photo by Brian Clark Howard)*

The CFL tube is filled with an inert gas, typically argon but sometimes neon, as well as a small amount of mercury vapor, all at low pressure. As with larger fluorescent tubes, excited mercury atoms produce UV light, which strikes the phosphor coating on the inside of the glass, which emits visible light.

In all fluorescents, the phosphor coating is a key part of the design and is constantly evolving. It is primarily this feature that governs light color and quality, and coatings are selected to balance cost, efficiency, and light characteristics. Today, most CFLs use a layering of two to three phosphors (you'll often see "triphosphor," for three, advertised). A number of brands advertise higher blends, however, such as BlueMax, which boasts of "a five-phosphor blend to produce a light rich in all colors of the spectrum." However, be aware that such terms aren't really regulated, and you're best off testing before buying.

Most CFLs have electronic ballasts, although some earlier models had magnetic ballasts, which could cause flickering and humming.

Integrated versus Nonintegrated CFLs

CFLs can be manufactured as integrated or nonintegrated units. Integrated lamps combine the tube and ballast into a single product and have either an Edison screw-type (standard light bulb twist) or bayonet fitting (which fastens by means of a male side with pins and a female receptor with matching L-slots and springs). These are the CFLs that are most familiar to consumers, and they allow simple replacement of incandescent bulbs (Figure 4-8). This lowers the cost of CFL use because people can reuse existing hardware. CFLs should

FIGURE 4-8 CFLs can work great in vintage fixtures such as the one pictured, as well as new hardware that is specifically designed for them. *(Photo by Brian Clark Howard)*

work with most standard fixtures, although some people have reported problems with very old units and wiring. Optimal results usually can be achieved with dedicated CFL fixtures (i.e., fixtures designed specifically for them). Special three-way, candelabra and dimmable CFL models are also available for use when those features are needed.

Nonintegrated CFLs have a separate, replaceable bulb and permanently installed ballast. Since the ballasts are placed in the light fixture, they are larger and last longer than the integrated ones. Nonintegrated CFL housings tend to be more expensive, costing anywhere from $85 to $200 for each recessed light fixture. If a ballast with dimming capabilities is desired, the cost is anywhere from $125 to $300 per fixture. Nonintegrated CFLs are used mostly in commercial settings, such as hotels and office buildings, where lighting gets heavy use.

Direct Current (dc) CFLs

Most fluorescents will work on dc as long as there is enough voltage to sustain an arc. However, unless the starting switch is arranged to reverse the polarity of the supply to the lamp each time it is used, the mercury will accumulate at one end of the tube. An alternate solution is to hook up an inverter before the fluorescent light, which will convert the power from dc to ac.

Alternately, look for a CFL that is predesigned to work on dc. These are available online and are popular for recreational vehicles and off-the-grid homes, which may be powered by solar panels, wind turbines, or other generators that produce dc. CFLs are also sometimes used in developing countries, where they can be powered directly by car batteries or small renewable sources, even hand cranks.

CFL Advantages

Most household CFLs operate on 13 to 25 watts of energy, far less than the typical 60 to 100 watts for incandescent bulbs. Yet CFLs produce the same amount of light. A quick comparison with incandescent lamps shows that CFLs:

- Use less electricity (roughly 75 percent less).
- Have a longer rated life (8 to 15 times, which is commonly estimated at 10 times). CFLs typically have a rated lifespan of 6,000 to 15,000 hours, whereas incandescent bulbs usually have a lifespan of 750 to 1,000 hours.
- Cost only a bit more up front but save you an average of $30 over the life of the bulb, according to the EPA (some writers estimate $47, the Environmental Working Group estimates $80, and New York City—where energy prices are high—says $100).
- Save 2,000 times their own weight in greenhouse gases (450 pounds), according to the EPA. They conserve much more energy than is embodied in making and shipping the bulbs themselves.

U.S. News and World Report recently estimated that a household that invests $90 in changing 30 fixtures to CFLs would save $440 to $1,500 over the life of the bulbs depending on the local cost of electricity. The magazine suggested looking at your utility bill and estimating a 12 percent discount if you switch to the technology.

It is true that CFLs produce a somewhat different type of light than incandescent lamps. However, it is important to note that advanced phosphor formulations have substantially improved the subjective color of the light emitted by CFLs such that the best "soft white" CFLs are quite close to standard incandescent lamps. At a recent test in the lighting booth of the Green Depot store in New York City, two shoppers could not differentiate between the light quality from a new soft white CFL and a standard incandescent bulb. However, when an older-style CFL was switched on, the shoppers immediately criticized the "harsh" bluish light. Two examples of "prettier" CFLs include the 15-watt Bright Effects CFL, at 2,644 K, and the 14-watt Sylvania CFL, at 3,000 K. Earlier CFLs had a color temperature in the 6,500 K range.

Concerns with CFLs

No technology is perfect, of course. While CFLs and fluorescents in general provide many benefits, there are some drawbacks that you should be aware of. Fortunately, there are also solutions to many of these problems.

"Harsh" or Unpleasant Light Quality

Many people have complained that they don't like the light from CFLs. This is partly because they didn't get a good impression from older models or because they used lower-quality brands. However, it also may be because they used a cooler CFL. Typical cool white CFLs have high color temperatures, often well above 4,100 K and a CRI of around 62 to 65. While this may be suitable for some tasks, it isn't the most flattering light.

In fact, some studies have linked this level of fluorescent lighting to some discomfort in test workers. Reported symptoms have included headaches and a jittery feeling, as well as increased levels of stress hormones.

However, there are several things to keep in mind. Recently introduced CFLs are much improved and are available with much warmer color temperatures and better color rendering. Some people also have found that full-spectrum and so-called high-definition CFLs, which have enhanced light in the blue wavelengths, help to reduce glare, eye strain, and fatigue, as well as provide a more natural, comfortable light. Full Spectrum Solutions, which makes Blue-Max high-definition CFLs, claims that its new lights help to keep the pupils of the eye small and stable, which is thought to improve vision quality and possibly mood.

Decreased Lifespans and Switching Frequency

One of the biggest complaints we've heard about CFLs is that they don't last as long as boosters suggest. This can be a bit of a complicated issue, and the lifetime of any lamp depends on a number of factors, including manufacturing quality, subtle defects, operating voltage, voltage spikes, ambient temperature, mechanical shock, stress from the elements, usage patterns, and frequency of cycling on and off.

Many people may not realize that the lifespan of a fluorescent drops significantly with frequent turning on and off. In fact, a number of lighting installers have warned that frequent switching in five-minute cycles can decrease the lifespan of a CFL up to 85 percent, making it about on par with an incandescent. Others have estimated that the average CFL can withstand about 7,000 on-off cycles.

This issue is further complicated by the folk wisdom that says fluorescents should be left on for long periods because they are commonly believed to use a surge of energy when they start up. This belief originated back in the middle of the twentieth century, when energy prices were cheap and lighting technologies were much less efficient. In fact, according to the DOE, it takes only about five seconds of use for a fluorescent light to use the same amount of power that it requires to start up. For an incandescent bulb, it takes even less time. However, since frequent switching does decrease the life of fluorescent technology, it is wise to strike some balance. The DOE suggests turning off fluorescents if you won't need them for 15 minutes or more. With this in mind, it's not a bad idea to install CFLs in areas where you won't be switching them on and off so often. One of us tried a CFL in a closet fixture and found that it lasted only about as long as an incandescent bulb because it was rapidly switched on and off. Even in such a worst-case scenario, however, you'll still save energy on your lighting, and CFLs now have become so cheap that using them up faster need not necessarily be as big a deal as critics make it out to be. (Learn to calculate just how much energy you can save by turning your lights off in Chapter 8.)

It's also worth pointing out that as with most lighting, the life of your bulb can be extended if you keep the fixture clean and dry and if the wiring is done correctly, with proper grounding.

Time to Achieve Full Brightness

Another common complaint about CFLs is that they don't always light up to full strength immediately, the way familiar incandescents do. According to the Environmental Defense Fund, some CFLs provide only 50 to 80 percent of their rated light output at "initial switch on" and can take up to about three minutes to warm up completely. Color cast also may be somewhat different at first. To meet the standard, Energy Star–rated CFLs must turn on in less than a second and reach at least 80 percent of full light output within three minutes.

In practice, warm-up time can vary considerably among brands and types, although newer models tend to switch on much faster. The brand-new CFLs tested at the Green Depot display lit up to full brightness immediately. Some users also have reported that the time lag can be worse at colder temperatures.

If this issue is a concern for you, you might try the "instant on" CFLs that have been marketed recently. You also might consider so-called cold-cathode CFLs, which reach their rated light output quicker (more on these below).

Light Decay

It's also true that CFLs produce less light as they age. The light output decay is exponential, with the fastest losses being soon after the lamp is first used. The good news, though, is that even by the end of their lives, good-quality CFLs should produce 70 to 80 percent of their original light output. For most people, such a small decay should not be noticeable, although it is good to be aware of in case you have specialized lighting needs.

Note that this hasn't always been the case. As we've learned, CFL technology has improved considerably over the past few years, and earlier versions did suffer from worse light decay. When the DOE looked at CFLs made in 2003 and 2004, it found that one-quarter of them no longer met their rated output after just 40 percent of their rated service life. Today this problem is less likely, thanks to improved manufacturing.

Audible Noise

Energy Star–qualified CFLs are not supposed to emit an audible noise, and most offerings on the market today are silent. In the past, some CFLs did emit a faint buzzing sound, which certainly can be annoying. This is rarely an issue today.

Trouble with Timers and Sensors

Electronic timers can interfere with the ballast in CFLs, although mechanical timers shouldn't be a problem. Some motion sensors and other electronic controls may not work with CFLs.

Heat Damage

Many CFLs have ballasts that are easily damaged by heat. This can include the heat that builds up in enclosed recessed lighting fixtures. However, the good news is that there are CFLs specifically designed for these purposes (they usually come in the reflector style).

Remember, too, that CFLs generate much less heat than incandescents, which means that they add less to your cooling loads for interior spaces (although it also means you'll have less heating assistance in winter).

Vibration Damage

Most CFLs should not be used in vibrating fixtures, such as on ceiling fans, unless they are specifically designed for that purpose. Vibrations can shorten their lifespan and make breakage more likely.

Interference with Infrared Signals

This is not common, but some people have reported that CFLs interfere with infrared (IR) remote operation of TVs, DVD players, cordless phones, cell phones, and other devices. It is possible for electronics systems to mistake the IR light that is emitted by a CFL for signals. However, *Consumer Reports* concluded that Energy Star–qualified CFLs are less likely to cause such interference. The magazine added that if a particular CFL is known to cause interference, it is supposed to carry a warning on the packaging.

If you do notice that CFLs cause interference, try moving them away from your device. Or try swapping out for another bulb.

Power Factor: Effects on Power Quality

CFLs behave a bit differently than incandescents when it comes to the electricity in the incoming lines. While this isn't normally something most consumers need to worry about, it can affect large-scale installations, and it is definitely something that utilities are concerned about. Fluorescent lighting behaves in a nonlinear way when it comes to ac power and thus, without corrections, has a lower *power factor*. Power factor essentially measures the load on the line, and something with a low power factor (<0.85) draws more current and is less efficient than a load with a high power factor (closer to one) for the same amount of useful power. In other words, powering devices with low power factors takes more current from the utility.

When Margery Conner of *EDN* (*Electronics Design, Strategy, News*) tested a typical CFL in her home with a Kill-O-Watt energy monitor, she recorded a power factor of .57. "This is lousy," she wrote. Conner then asked Peter Banwell of the EPA's Energy Star Program if the agency had considered the issue. "We looked at this in detail several years ago and decided against it, though there are a couple of utilities that still support the idea," Banwell told her. "We may take this up in the future, as the market share grows, but right now it is still in the noise in terms of impacts."

Conner concluded that the relatively low power factor of CFLs shouldn't affect home owner electricity bills, and she stressed that the overwhelming efficiency of the bulbs more than makes up for any small inefficiencies caused by the load issue. She questioned whether it would be worth it to add corrective circuitry to CFL ballasts because this would increase the price of the bulbs.

Across the grid, it's clear that the benefits of installing CFLs, in terms of energy savings, far outweigh any small inefficiencies relating to power factor (Figure 4-9). Still, the issue is significant enough for green products certifier Green Seal to call for improving and standardizing the power factor of CFLs.

A similar electrical concept is harmonic distortion, and it's true that CFLs can "distort" the electricity on the line. Again, this isn't

FIGURE 4-9 CFLs can work with almost any décor. *(Photo by Brian Clark Howard)*

something typical users need to worry about, although some engineers have pointed out that it could be an issue in closed systems that are, say, running solely on solar power. It's also worth paying attention to if you are installing a large number of CFLs at a big facility, because utilities often charge heavy users for low power factors. One important point is that you can buy CFLs that have low (below 30 percent) total harmonic distortion (THD) and power factors greater than 0.9.

Fluorescent Lamps and Mercury

One can hardly talk about CFLs without being confronted with the issue of mercury content. After all, they did descend from something called *mercury vapor lights*. There is considerable fear in the marketplace about the technology as a result, and no wonder, because mercury is a persistent toxin that inhibits childhood development and can impair the nervous system.

At least one green lighting installer avoids selling CFLs because of the mercury issue. Lenny Gianfrancisco of Solar Science Group and the American Green Energy Council in Port Charlotte, Florida, estimates that one billion fluorescent lamps are disposed of every year globally. Gianfrancisco says that amounts to 50,000 pounds of mercury waste, which he says is theoretically enough to pollute every gallon of water in the United States and Canada. (Gianfrancisco pushes LEDs as safer alternatives, which we'll get to in Chapter 5.)

The staff of *The Daily Green* has received numerous calls and e-mails from consumers who are worried about mercury in CFLs. One reader said that a local hazardous waste company had visited her neighbor after a CFL was broken in her home. The contractors allegedly arrived in full hazmat gear, looking like astronauts, and supposedly did a lengthy "detox" of the home, leaving a bill for hundreds of dollars. *The Daily Green* editors assured the woman that her friend had been ripped off, assuming the story is true (there have been urban legends to this effect that haven't always checked out).

Why is such a reaction overkill? Well, there are several important points to keep in mind. For one thing, CFLs contain an exceedingly small amount of mercury. This is, on average, only about five milligrams, although some sources now put the average around four milligrams because manufacturers have been steadily decreasing the amount. There are currently a number of CFLs on the market with as little as one milligram of mercury, which we'll get into shortly. To put this in perspective, five milligrams is roughly the amount of mercury that would fill the period at the end of this sentence. Compare that with the 25 milligrams of mercury in a typical watch battery, the 500 milligrams in an old fever thermometer, and the 3,000 milligrams in a manual home thermostat.

There is also a comparable amount of mercury in other types of gas-discharge lighting, such as is commonly found in gymnasiums, as well as in switches in auto parts and in various computer components, not to mention dentists' offices. Further, old-fashioned tube fluorescent lights are bigger and contain more mercury. While there is indeed pressure from environmental groups on manufacturers to decrease the mercury content in these other products, there isn't as much concern from the general public (with the possible exception of mercury amalgam fillings for teeth). Yet how many people do you know who are afraid to use a computer or work in an office because of concerns about mercury contamination? Perhaps this is so because we have an apparent need to make every "green" product as clean and perfect as possible, right off the bat.

Low-Mercury and Safety CFLs

The industry continues to work on decreasing the amount of mercury in CFLs. The average four-foot lamp made in 1995 contained 75 percent less mercury than one manufactured in 1985, and the amounts keep dropping. Today, a number of manufacturers make CFLs with 70 percent less mercury than typical CFLs. Examples include Alto from Philips and Ecobulb Plus from Feit Electric.

In December 2008, Energy Star set a maximum mercury standard of five milligrams for CFLs that carry its label. However, as the Environmental Working Group (EWG) points out, U.S. manufacturers had voluntarily set that mercury level as a target the year before. Even so, when the EWG tested CFLs on the market in 2009, it found considerable inconsistency. Several Energy Star–registered CFLs had too much mercury to be legal for sale in Europe, where the European Union caps the content at four milligrams per CFL. On the other hand, seven bulbs, from brands Earthmate, Litetronics,

Sylvania, Feit, MaxLite, and Philips, had significantly less mercury (Table 4-1). These bulbs had one to 2.7 milligrams of mercury, and all lasted 8,000 to 15,000 hours, longer than the Energy Star standard of 6,000 hours. All seven also had higher energy efficiency.

TABLE 4-1   CFLs with Low Mercury Contents, According to Tests by the Environmental Working Group *(Environmental Working Group)*

Brand and Bulb Line	Mercury per Bulb	Average Life Span	Where to Buy
Earthmate Mini-Size Bulbs (13, 15, 20, and 23 Watts)	About 1 mg	10,000 hours	Energy Federation
Litetronics Neolite (10, 13, 15, 20, and 23 Watts)	About 1 mg	10,000 hours	1000bulbs.com
Sylvania Micro-Mini (13, 20, and 23 Watts)	Less than 1.5 mg	12,000 hours	Amazon.com
Sylvania DURA-ONE Mart (reflector bulbs)	Less than 1.8 mg	15,000 hours	Conservation
Feit Ecobulb	Less than 2.5 mg	8–10,000 hours	Amazon.com
MaxLite	1.2–2.5 mg	10,000 hours	Amazon.com
Philips with Alto	1.23–2.7 mg	8–10,000 hours	blackEnergy

Source: http://www.ewg.org/node/27221

Amalgam-based CFLs

T. J. (Tom) Irvine of southern Florida got into the lighting business 12 years ago because he got sick of changing burned-out light bulbs and figured that there had to be a better solution. Today, Irvine is president and CEO of TAG Industries, Inc., a manufacturer of Clear-Lite fluorescents and other efficient lighting products. In a recent interview, Irvine said that he was first focused on getting the best light possible out of CFLs, especially for doing detail-oriented work. The result is his ClearLite "reading and craft" CFLs, which he is convinced help people "see better."

Then one of Irvine's young children knocked over a lamp in his room, shattering the CFL. As a parent, Irvine panicked, worrying about mercury exposure to his family. This spurred him to take a closer look at the toxic element. Irvine's brands had already been using mercury in *amalgam* form, meaning that it is mixed with other metals (this is how it appears in tooth fillings, mixed with silver, copper, tin, and zinc). According to Irvine, amalgam is much safer for factory workers to handle, resulting in fewer cases of poisoning, and it is easier to control, making sure that no more mercury than necessary is loaded into each device. "When the bulb is off, the mercury is bound up safely in the amalgam," said Irvine. "When it is on, the mercury vapor is released into the tube so that it can work." (True, if a bulb is lit when it shatters, much of the mercury will be present in vapor form, which is by far the most dangerous.)

In addition to all TAG Industries' products, many lighting manufacturers are using mercury in amalgam form, including TCP, Litetronics, Sylvania, and Philips, which also argue that the technology allows for more even lighting and better flexibility to temperature ranges.

Irvine and his engineers spent six years looking for other ways to reduce mercury risk. The result was the ArmorLite CFL, in which the bulb's spiral is contained within a silicone composite sealed shell. Irvine said that this shell should contain any mercury if the inside glass becomes compromised. He said that he has tested the bulbs under normal breakage conditions, and analysis by the independent Cambridge Labs could find no measurable mercury levels outside the outer dome (Figure 4-10).

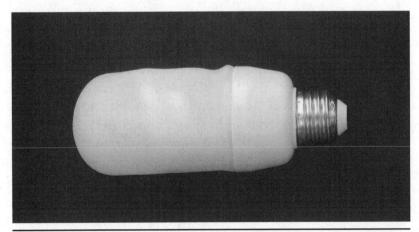

FIGURE 4-10   Even though this ArmorLite bulb was broken, all the contents—including the toxic mercury—remain inside the silicone safety shell. *(ClearLite)*

ArmorLite CFLs are approved by the Food and Drug Administration (FDA) as covered lighting to use around food, and they are UL listed for damp locations. "ArmorLite allows the opportunity to leave a bulb wherever you need it, be that in a child's room, or in the room of someone who is elderly or immune compromised, which are the people the EPA tells us are most at risk from mercury exposure," Irvine said. "My goal was an efficient lighting technology that is also more earth-friendly," added Irvine. "I liken it to diesel engines, which were always more efficient, but haven't always been earth-friendly."

If you are wondering if the silicone shell blocks too much of the output of an ArmorLite bulb, Irvine countered that it actually produces a softer, more even light. Moreover, the bulbs more closely resemble the incandescents we're all used to. At $7 to $15 online, ArmorLite costs about the same as the best regular CFLs.

CFLs Actually Reduce the Amount of Mercury in the Air

It's also important to realize that CFLs actually result in less mercury released into the environment than standard incandescents. This may seem surprising, but think about it a minute. The number one source of mercury in our environment comes from burning fossil fuels, especially coal, which is the most common fuel in the United States. Since CFLs are so much more efficient than incandescents, a power plant actually will emit 10 milligrams of mercury to produce the electricity to run an incandescent bulb compared with only 2.4 milligrams of mercury to run an equivalent CFL for the same amount of time.

Another way to look at this is that even if all 270 to 290 million CFLs sold in the banner year of 2007 were sent to landfills, this still would represent only 0.1 percent of all U.S. emissions of the heavy metal for that year (around 0.13 ton of the total 104 tons), according to the Natural Resources Defense Council. In this sense, therefore, using CFLs helps us to avoid more mercury pollution because the real culprit is burning coal (this is also what puts mercury in the oceans where it bioaccumulates in the fish we eat, resulting in a serious health concern for those who eat seafood).

Also, it's worth remembering that the more energy we use, the more we have to drill, mine, refine, and transport fuels. All this takes a considerable toll on the environment, not to mention increasing the risks of spills, explosions, and other problems. The less we exploit natural resources, the less mercury we pull up from the crust.

But What Happens if a CFL Does Break?

Don't panic! According to Helen Suh MacIntosh, a professor of environmental health at Harvard University, exposure to mercury from a broken CFL is unlikely to cause any harm. Even if all the bulb's mercury vaporizes into the air on breakage, which is unlikely because most of it typically adheres to the phosphors, Suh MacIntosh says that the concentration in a room still should be lower than Occupational Safety and Health Administration (OSHA) safe standards.

However, there is still sufficient concern and uncertainly about possible exposures, especially when it comes to children, the elderly, and the immune-compromised. When the Maine Department of Environmental Protection did lab tests on broken CFLs, it did find mercury levels that it considered to be unsafe. However, the Lawrence Berkeley National Laboratory reviewed the Maine data and concluded that "the most extreme CFL breakage scenario measured in [the] Maine study only equaled the approximate exposure from a single meal of fish."

As the scientific debate continues, it's probably prudent to take any broken CFLs seriously. Here's what you should do, according to the EPA. Actually, first let's look at what *not* to do:

- Don't use a vacuum cleaner or broom, which can spread mercury around.
- Don't pour mercury down the drain because it can damage plumbing and harms the water supply.
- If powder or glass from the bulb comes in contact with clothing, discard the clothing. It's better not to try to launder it because this could spread the mercury around.

Now, what you *should* do:

- Keep pets, children, and pregnant or nursing women away from the area until it is cleaned.
- Open windows, turn on fans, and leave the room until it is ventilated for 15 minutes. If you have central air conditioning, turn it off.

- The EWG recommends donning rubber gloves, safety glasses, and a dust mask, although the EPA does not explicitly recommend these steps.
- Carefully scoop up any bulb fragments and powder with a piece of stiff paper or cardboard. Make sure that nothing touches any bare skin.
- Place the fragments in a sealable glass jar (best), or if you don't have a jar, double bag the fragments in sealed plastic bags (although note that a recent test by the Maine Department of Environmental Protection measured mercury vapor escaping from a sealed plastic bag, and the EPA is currently investigating this).
- Use sticky tape, such as duct tape, to pick up any remaining powder and fragments.
- Wipe the area clean with damp towels, and put the towels in the jar or bag. If you wore rubber gloves, place the gloves in the jar or bag too.
- If the spill occurred on carpet or a rug, the EPA says that you can vacuum the area after you complete the preceding steps. However, if you do, take out the vacuum bag when you are done, and place that in the jar or bag as well (if possible).
- Take all these materials to your recycling center or hazardous waste dump (more on this below).
- The next few times you vacuum your house, let it ventilate before and after you clean, and turn off your central air conditioning during the process.

Don't know where to take broken or unwanted fluorescent bulbs? Don't worry, a proper disposal site should be easy to find. You can look one up on lamprecycle.org, which has state and local listings of qualified sites and is run by the National Electrical Manufacturers Association. Or call your local town hall or look on your town's Web site. The listing could be under "recycling," "hazardous waste," "disposal," or "transfer station." Your town dump and/or recyclers should be able to help you, or possibly even a fire station. Many towns have dedicated dropoff places that accept fluorescent lights. In some cases, you may have to save it until a certain day of the week or month.

Some retailers are also making it easier for consumers to dispose of spent CFLs. Home Depot, for example, has sponsored a recycling program. In Minnesota, consumers have the option to drop off CFLs at hundreds of participating retailers. A similar program has been offered by Sears stores in Indiana and by various partners in the Pacific Northwest.

At proper facilities, spent fluorescents are safely broken down, often by machines called *Bulb Eaters* (Figure 4-11). A Bulb Eater crushes the lamps inside 55-gallon drums, while safely extracting the mercury. When the drum is full, it is sent to a recycling facility with a *retorting machine*, which purifies the mercury. The valuable element can then be resold. The leftover aluminum from the lamps gets sold as scrap, where it makes its way into a wide range of products. The glass typically is used as aggregate for concrete or fiberglass.

Note that it is now illegal in many places to toss fluorescent lights in the regular trash, although enforcement tends to be spotty

Figure 4-11 A "Bulb Eater" for safely crushing fluorescent lights for recycling. *(Air Cycle Corporation)*

at best. The good news is that most landfills are supposed to be lined, and mercury placed in them theoretically should be contained. As it stands now, 70 percent of all heavy metals entering U.S. landfills come from electronics waste, according to the Silicon Valley Toxics Coalition, and lighting is a much smaller contribution. In Europe, producer takeback laws mandate that bulb manufacturers provide for safe disposal of the products, and a recycling fee is built into the price of the item. This isn't the case yet in the United States (although we could ask Congress for it!). As a result, some critics have estimated that barely three percent of CFLs get disposed of properly in the United States, although a recent *Wall Street Journal* report estimated that roughly 25 percent of mercury-containing bulbs do get recycled at licensed facilities, thanks in large part to the actions of commercial facilities.

If taking your old CFL someplace sounds like a hassle, there's another option, too. You can order a RECYCLEPAK from bulb maker Sylvania. It's a handy kit, with prepaid shipping, that allows you to send in spent bulbs for proper disposal (Figure 4-12). You'll have to pay a small fee for this, but take heart, recycling costs amount to just about one percent of the total amount of money you'll spend on a bulb in its lifetime because the energy needed to light it takes up the lion's share.

Figure 4-12 A RECYCLEPAK from Osram Sylvania makes it easy to return spent CFLs for safe disposal. *(Osram Sylvania, Inc. and Veolia ES Technical Solutions)*

When a CFL Stops Working, Will It Start Emitting Mercury?

This shouldn't happen because fluorescents are designed to be self-protective at the end of their useful lives. It's also possible that a ballast could fail, but again, this should not result in release of mercury (some release of smoke is possible, though uncommon).

How to Further Minimize the Risk from Mercury in CFLs

Choose CFLs for locations that are less likely to result in breakage, such as in high fixtures that don't undergo vibrations. Avoid putting CFLs in clip lamps or table lamps that get jostled around, such as in high-traffic areas. The EWG suggests avoiding CFLs in children's rooms or play rooms or near irreplaceable art or fabrics. However, you may want to consider using ArmorLite CFLs in such places.

One final point about mercury in CFLs is that there have been reports of workers in China who make the bulbs getting poisoned by the toxic element. This is clearly a serious concern, but unfortunately, the problem isn't limited to any one type of technology. As China has been rapidly expanding its industrial output, safety and environmental guidelines haven't always kept pace. Those who observe the sector hope that pressure from the West, as well as internally, will help Chinese suppliers to improve conditions.

Types of CFLs

CFLs are designed to withstand different conditions (Figure 4-13). Some of the most basic specifications include models that are built for the following conditions:

- Indoor
- Outdoor
- Indoor/outdoor
- Weatherproof

In general, CFLs that aren't designed for outdoor use have a hard time starting in cold weather. However, CFLs are available with cold-weather ballasts, which may be rated to as low as –23°C (–10°F). Bulbs rated as "weatherproof" are designed to stand up better to the

FIGURE 4-13   Some CFLs are specifically designed for outside use. *(Photo by Brian Clark Howard)*

elements. Note that cold-cathode CFLs (more below) tend to do better in a wide range of temperatures.

Standard CFLs typically are offered in these common wattages:

- 13 watts (60-watt incandescent equivalent)
- 18 watts (75-watt incandescent equivalent)
- 26 watts (100-watt incandescent equivalent)

CFLs are also produced in a range of colors for special applications. These include:

- Red, green, orange, and pink—primarily for novelty purposes
- Blue—used in phototherapy
- Yellow—for outdoor lighting (because it is believed to be less enticing to insects, although some observers have questioned the effectiveness)

- Black light (UV light)—for special effects (Notably, black light CFLs, which have a UVA-generating phosphor, are much more efficient than incandescent black light lamps because the amount of UV light an incandescent lamp produces is only a fraction of the generated spectrum.)

Got a gorgeous vintage sconce you need filling or a funky old lamp you don't want to part with? No problem! There are many flavors of CFL on the market today to fit nearly every fixture (Figure 4-14). Examples include:

- Classic spiral CFLs
- Tubular CFLs, which are slightly more efficient and which are more popular in Europe
- Covered CFLs, to mimic the look of traditional incandescents (A-line) and to work with harp shades that fit directly on the bulb
- Candelabra CFLs, to fit in large and medium fixtures, such as chandeliers and pendants (Many have adaptable bases that work in regular sizes or smaller candelabra holders.)
- Flicker flame-tip CFLs, which have an outer casing that looks like a previous era
- Minispirals, which fit in smaller fixtures and look cute to boot
- Track lighting CFLs, in sizes such as MR16 and GU10, which can replace halogen bulbs and have pin connectors
- Globe CFLs, for use in vanities (Some are made with waterproof housings.)
- Outdoor post CFLs
- Floodlight CFLs
- Reflector CFLs
- Three-way CFLs (One example we found at Walmart uses 12/23/29 watts and replaces a 50/100/150-watt incandescent three-way.)

Note that most CFLs with alternative shapes usually consist of spirals that are enclosed in a second layer of glass or plastic. This can add to the cost of the bulb and can extend the warm-up time somewhat. It also will make the bulb a bit less bright, although it can help to soften the light. It also may decrease the lifespan somewhat.

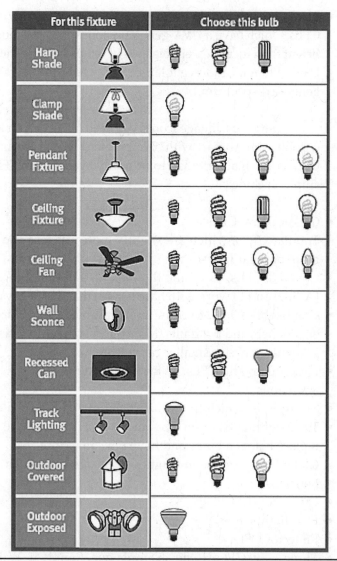

For this fixture		Choose this bulb			
Harp Shade					
Clamp Shade					
Pendant Fixture					
Ceiling Fixture					
Ceiling Fan					
Wall Sconce					
Recessed Can					
Track Lighting					
Outdoor Covered					
Outdoor Exposed					

FIGURE 4-14   A look at some of the different types of CFLs now available, with some suggested applications. *(Natural Resources Canada)*

CFLs also come in a variety of light tones, although note that considerable variability exists among manufacturers, and the colors aren't currently standardized. Your best bet is to stick with major manufacturers such as GE, Sylvania, Litetronics, and Philips or boutique providers such as ArmorLite and to try before you buy. Some examples include:

- Full spectrum (theoretically covering the entire range of visible light, 400 to 700 nanometers)
- Bright white (cooler)
- Daylight (cooler)
- Soft white (warmer)
- High definition (said to be full spectrum plus enhanced blues, which some say improves mood and visual acuity)

It is important to note that none of these terms are regulated when it comes to CFLs, and there is considerable variation between models and manufacturers. Some bulbs labeled "soft white" are indistinguishable from "daylight" bulbs from other companies. This can make it confusing and difficult for consumers to find a bulb they like, and there is no easy way around this issue, save for trying before you buy. (However, new U.S. regulations expected to go in effect in 2011 should make it easier, by requiring bulbs to list their color temperature, as well as other core data.)

Tom Irvine of TAG Industries and ArmorLight is particularly skeptical of manufacturer claims for full spectrum. "I always tell people to remember that it's still artificial light, and there is no substitute for the sun," he said. "We have spent years working on it, and you get closer when you use our two-bulb systems, which pair one of our 'warm and cozy bulbs' with one of our 'reading and craft' bulbs, and with both on, you increase the color spectrum and get brighter, more natural light."

Energy Star CFLs

Whenever possible, choose Energy Star–certified CFLs. They must come with a two-year warranty, they must have a rated lifespan of at least 6,000 hours, they cannot emit audible noise, and they must meet minimum requirements for energy efficiency and light quality. Energy Star CFLs also must turn on in less than one second and reach at least 80 percent of full light output within three minutes. As stated earlier, they must not have more than five milligrams of mercury.

Dimmable CFLs

Dimming definitely has not been a strong point for CFLs. Standard CFLs are not dimmable, and you should never put them on a dimming circuit. If you do, it will shorten bulb life and will void some warranties. It also could cause the ballast to heat up to the point of a fire hazard. Even if you always leave the dimmer on full brightness, it still could damage the bulb.

There are, however, a number of dimmable CFLs on the market today. If you purchase one, follow the instructions with the product, although note that most should work with modern dimmers. It is possible to get dimmer circuitry that is specifically designed for CFLs. For larger lights, it is often best to get dimming ballasts, which generally do a better, smoother job at dimming than placing a CFL on a traditional dimmer switch. Even so, CFLs haven't been that good at replicating dimmable incandescents. It's easy for incandescents to dim smoothly from 100 to 0 percent brightness simply by restricting the power going into the bulb. CFLs, however, don't function in the same way.

Dimmable CFLs darken in stages and often can't go below 20 percent illumination, at which point they either turn off or flicker. Similarly, they usually jump from about 100 percent brightness to 80 percent without gradations. And multiple dimmable CFLs on the same circuit don't always dim equally. Once you turn off the bulb, you'll have to start back at full brightness next time, unlike incandescents.

Perhaps underscoring the rapid progression of green lighting, dimmable CFLs are getting better all the time. Some manufacturers now advertise dimmable CFLs that darken to lower than 10 percent and brighten to close to 100 percent. However, designers who have tried these CFLs question those claims. Some dimmable CFLs are designed to work with regular light switches, and the ballast senses the number of times the switch is flicked in rapid succession in order to adjust dimming.

Another issue has been that when some CFLs dim, the color temperature stays the same, unlike with incandescents or sunlight, in which color gets warmer as the light gets dimmer. Some emotional response testing has suggested that people find dim bluish light to be cold or even sinister. Thus it's a good idea to get warmer-light CFLs if you are considering dimming.

Dimmable CFLs cost more than standard CFLs, and they haven't earned a very good reputation owing to current and past limitations. On a recent trip to the lighting district in New York City, a customer asked a sales representative, "CFLs can't be dimmed, right?" The salesman answered, "No. You don't want CFLs in your home." It's possible the salesman was trying to steer the customer toward more expensive products, but it's also possible that he simply did not know that dimmable CFLs have been improving. Green lighting designer David Bergman of New York, for example, gets attractive results in his Fire & Water collection with dimmable CFLs.

"People are likely to be more disappointed than not when it comes to current dimmable CFLs, although we are working on new ballasts," argued Irvine, who recommends energy-saving halogens if dimming is really important for a particular feature.

Fluorescent Induction Lamps

Lighting without electrodes was invented by Nikola Tesla, who demonstrated that he could illuminate incandescent and fluorescent lamps by wirelessly applying electromagnetic fields. Electrodeless fluorescent lamps have been available commercially since 1990, although they are poorly known by the general public. Instead of passing between electrodes, a current is induced in the gas column through electromagnetic induction. Since there are no electrodes to wear out, these lamps can last much longer, hundreds of thousands of hours in fact.

Induction lights do have a higher purchase price, however, but they are slowly being adopted in some commercial settings.

Cold-Cathode Fluorescent Lamps (CCFLs)

CCFLs have been marketed only recently as common lighting options. Cold-cathode fluorescent technology itself has been around for years and has been used as backlighting for liquid-crystal displays (LCD), in document copiers, and even in cell phones. This technology is also employed commonly by amateur computer case "modders," who build elaborate housings for their machines with working lights and special effects (often to mimic styles or devices from science fiction).

The name can be a bit misleading because it doesn't mean that the electrons in the lamp are necessarily cold. Actually, they can be quite hot, but they are liberated only by the level of potential difference provided by the electrodes, not by the additional presence of heat, as occurs in standard fluorescents. In technical terms, standard fluorescents operate at their thermionic emission temperatures, whereas CCFLs operate below this level (Figure 4-15). Since CCFLs need no thermionic emission coating, they can last much longer than regular fluorescents because it is the depletion of this coating that results in many bulb failures.

CCFLs have an efficacy (lumens per watt) that's about half that of standard CFLs. A typical CCFL bulb may use five watts of electricity to replace a 20- to 25-watt incandescent light bulb. CCFLs operate at a voltage that's about five times higher than that of CFLs, but the current is about 10 times lower. Since CCFLs have much smaller spirals, they require less mercury (roughly one-tenth).

Cold cathode lights can be fashioned into long-tube form, the lettering shape of signs, or into CCFLs. These days, CCFLs usually have an inner spiral shape inside an outer dome, which can be A shaped like traditional light bulbs, flicker-tip shaped for cande-

FIGURE 4-15   A CCFL. *(Litetronics)*

labras, or reflector shaped for recessed lighting. Colored holiday CCFLs are also available, as well as other niche applications.

Advantages of CCFLs

- CCFLs have a long life, with 50,000 hours sometimes cited. Typical CCFLs on the market are sold with a rated life of 18,000 or 25,000 hours.
- Longevity is not shortened by repeatedly turning them on and off, unlike standard fluorescents.
- CCFLs stay cool.
- CCFLs light up instantly, and they are flicker-free.
- CCFLs are dimmable and work with sensors, photocells, and other electronics.
- CCFLs have strong color rendering, with a typical CRI of 82.
- CCFLs work better in hot and cold environments than regular CFLs.
- CCFLs can be used as flashers.

Disadvantages of CCFLs

- CCFLs currently cost a bit more than regular CFLs ($13 to $35 versus $2 to $8), but they are cheaper than light-emitting diodes (LEDs).
- According to Tom Irvine of TAG Industries, CCFLs start "losing performance" at sizes above 15 watts, so he doesn't recommend them for applications that need very bright light.

Summary

Here are some tips for getting the most out of CFLs:

- Most CFLs work in common incandescent fixtures, but some may have trouble operating in older fixtures.
- If you are dealing with new construction or upgrading very old fixtures, consider installing dedicated fluorescent fixtures—they allow higher energy savings and better light, reliability and lifespan.
- Start with high-use fixtures that aren't switched on and off frequently, and replace those with CFLs.

- Avoid fixtures with a high risk of breakage, near irreplaceable art or fabrics, or in children's rooms. Or consider newer safety CFLs.

While CFLs do contain toxic mercury, the amount is dropping, and there are a number of strategies to minimize risk. Overall, most experts say that the benefits of using fluorescents far outweigh the dangers and negatives. The technology has shown impressive improvements and is much more versatile than before. For many applications, CFLs and fluorescent tubes are the most practical, affordable green lighting currently available.

CHAPTER 5

Light-Emitting Diodes (LEDs)

In November 2009, 2,076 eleven-watt incandescent bulbs were removed from the iconic Reno Arch in Reno, Nevada, and replaced with more energy-efficient 2.5-watt light-emitting diodes (LEDs). This $62,180 retrofit is expected to reduce electricity demand by 92,011 kilowatts, saving $10,441 per year for "The Biggest Little City in the World." The upgrade is part of a citywide energy and water efficiency plan that is projected to save $1 million a year and create 222 jobs. At the lighting ceremony, the old incandescents were passed around the crowd as souvenirs of the past.

In California, Los Angeles International Airport (LAX) officials switched to LEDs for functional and outdoor lighting, reducing annual lighting costs by $55,000 and lifetime maintenance costs by $980,000. In Oconomowoc, Wisconsin, Sentry Equipment Corporation chose to light its new factory almost entirely with LEDs, both interior and exterior. The initial cost was three times more than incandescent and fluorescent bulbs, but this price premium is expected to be repaid within two years from electricity savings. The bulbs are expected to last for 20 years.

In New York's Times Square, the world-famous ball that descends every New Year's Eve is now covered with bright, efficient LEDs. In Las Vegas, thousands of tourists have been delighted by the 1,500-foot-long shimmering LED display of the Fremont Street Experience (Figure 5-1). In addition, thousands of LEDs brought

FIGURE 5-1 The Fremont Street Experience is a massive LED display over the heart of Las Vegas. Is it a symbol for the evolution of lighting? *(Photo by Brian Clark Howard)*

light, color, and movement to the opening ceremony of the Beijing Summer Olympics.

On the Rhode Island coast, Joe Hageman and Kim Lancaster's 70 recessed LEDs save 7,730 kilowatt-hours over standard incandescents and provide excellent, cozy light in their family home.

"LEDs are the most advanced lighting technology we have now, and people are excited about it," Avani Pavasia explained in the New Generation Lighting shop on the Bowery in New York City (in the lighting district). Pavasia is a young lighting and interior designer who hopes to work on green projects. "LEDs are especially being used now for commercial work, because that sector is more aware. Typical consumers aren't that educated about them yet," she added.

Pavasia pointed to a massive chandelier on the ceiling of the store. An array of blue LEDs shimmered through the etched glass. "That can change to many colors, since there is a separate controller," Pavasia explained.

So LEDs are future technology, here now. A recent study by Pike Research predicts that LEDs will account for nearly half of a $4.4 billion market for lamps in the commercial and industrial sectors by 2020, and residential applications are rapidly expanding. General Electric (GE) is now spending half its lighting research budget on the technology. Yes, LEDs are still more expensive than conventional lighting, but their cost is falling rapidly, their quality is improving by leaps and bounds, and they can save you serious energy right now. Some advantages of LEDs include:

- LEDs typically use 90 to 95 percent less energy than incandescent bulbs.
- LEDs are expected to last 10 to 20 years and provide tens of thousands of hours of light (40,000 to 50,000 hours is commonly estimated, although some models are reputed to last 100,000 hours. At 50,000 hours, that's 50 times longer than an incandescent).
- LEDs come in many colors, sizes, styles, and fixtures. They can produce different colors more efficiently than other lighting technologies.
- LEDs don't produce as much heat as other lighting, so they have reduced fire risk and less heat gain in interior spaces.
- LEDs are extremely rugged and are well suited to tough environments, such as on vibrating equipment or in extreme conditions.
- LEDs can be switched on and off very rapidly, faster than other lighting technologies. Many also can be dimmed.
- LEDs achieve full brightness in microseconds, much faster than fluorescents and about 10 times faster than incandescents.
- LEDs can be made extremely small, giving them many applications.
- LEDs do not contain toxic mercury.

History and Technological Overview

LEDs are a type of *solid-state lighting* (SSL), which means that they are built entirely from solid, nonmoving parts that conduct electrons. Solid-state technology therefore is distinct from vacuum-tube and gas-discharge technologies and includes such components as transistors and microprocessors. To create more light, LEDs can be clustered into "bulb" housings or "strip" housings. They work well on low-voltage circuits and can be made to work with line voltage (more on this later). As with all lighting, the LEDs available on the market can vary widely in light output and efficacy.

Although LED lighting may seem cutting-edge, the underlying technology essentially has been known for 100 years. Back in 1907, a British researcher named Henry Round was working at Marconi Labs when he noticed light being emitted by a semiconductor diode. Then, in the mid-1920s, a brilliant Russian scientist by the name of Oleg Vladimirovich Losev independently observed the same effect (Figure 5-2).

In 1955, Rubin Braunstein of the Radio Corporation of America reported on infrared emission from diode structures made with gallium antimonide (GaSb), gallium arsenide (GaAs), indium phosphide (InP), and silicon germanium (SiGe) alloys. According to the Smithsonian Institute, in 1961, Robert Biard and Gary Pittman of Texas Instruments received a patent for a GaAs infrared LED.

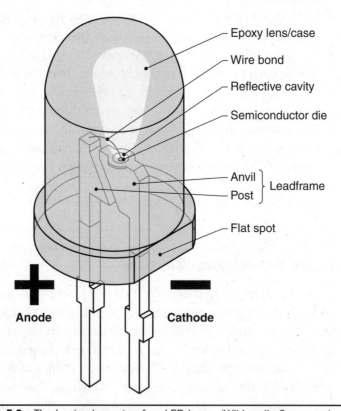

FIGURE 5-2 The basic elements of an LED lamp. *(Wikimedia Commons)*

The first practical visible-light LED was invented the following year by Nick Holonyak, Jr., of GE. Holonyak therefore is sometimes called the "father of the light-emitting diode." For the next few years, LEDs were extremely expensive, so few were produced. In 1968, the Monsanto Corporation found a way to mass produce visible-light LEDs for indicator lights using gallium arsenide phosphide. These were used in sophisticated laboratory equipment, although they soon started showing up in consumer electronics, starting with Hewlett-Packard (HP) calculators and then TVs, radios, and many other products.

Over the next few years, other colors of LEDs were developed, starting with yellow and green. Bright blue LEDs were invented in the 1990s by Japanese scientist Shuji Nakamura of Nichia Corporation. This was considered a major breakthrough because it allowed mixing of the primary colors to create a palette of 16 million colors and because bright blue LEDs became the stepping-stone to bright white LEDs, which can be used for lighting. Before that, LEDs were only bright enough to be used as indicator lights. Nakamura is sometimes called the "Thomas Edison of the LED industry," and he is the subject of the book *Brilliant! Shuji Nakamura and the Revolution in Lighting Technology*, by Bob Johnstone (Prometheus Books, 2007). Today, Nakamura continues to research the next generation of LEDs, as well as solar cells, as a professor at the University of California, Santa Barbara, and as a consultant to major LED maker Seoul Semiconductor.

LEDs have seen their efficiency and light output increase exponentially, with a doubling occurring about every 36 months since the 1960s. This trend is sometimes called *Haitz's law*, after Dr. Roland Haitz, a now-retired scientist from Agilent Technologies. Haitz's law states that every decade, the cost per lumen of light emitted falls by a factor of 10 and the amount of light generated per LED package increases by a factor of 20 for a given wavelength of light. This is similar to the more-famous *Moore's law* of computer processing power, which states that the number of transistors in a given integrated circuit doubles every 18 to 24 months. Since LEDs and computer chips are both solid-state electronics, it's not surprising that they would have parallel trajectories. Of course, LEDs also have benefited from advancements in materials science and optics.

How Does an LED Work?

An LED is a type of semiconductor diode. What's a diode? It's a two-terminal electronic component that conducts electric current in only one direction. In an LED, a chip of semiconducting material is impregnated, or doped, with impurities to create a structure called a *p-n junction*. These junctions are fundamental building blocks of many electronics, including transistors, integrated circuits, and solar cells. As in other diodes, current flows easily from the *p* side (positively charged), or anode, to the *n* side (negatively charged), or cathode, but not in the reverse direction. Electrons and electron holes flow into the junction, and when an electron meets a hole, it drops to a lower energy level, thereby releasing a photon. This effect is called *electroluminescence* (Figure 5-3).

The actual LED is quite small, usually less than one square millimeter. Additional optical components are added to shape and direct the light. LEDs are made from a number of inorganic semiconductor materials, many of which produce different colors of light. These include gallium arsenide, gallium phosphide, gallium indium phosphide, silicon carbide, sapphire, zinc selenide, and others. Recently,

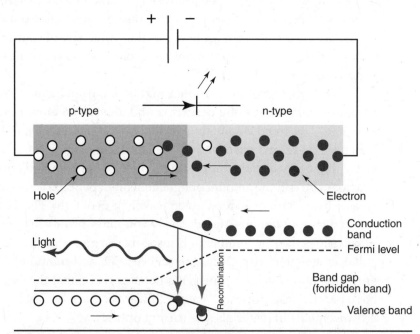

FIGURE 5-3   A representation of how LEDs create light. *(Wikimedia Commons)*

a dozen major lighting manufacturers banded together to create industry standards for LED light engines under a working group called *Zhaga*. Their efforts should help to streamline development.

The efficiency of LEDs has risen sharply and is currently up to around 130 lumens per watt in the laboratory and in some products available on the market (although more typical LEDs average 32 to 50 lumens per watt). Compare this with a 60- to 100-watt incandescent bulb, which produces only 15 lumens per watt, or a fluorescent lamp, which produces up to 100 lumens per watt. However, one issue with LEDs is that efficiency drops dramatically with increased current (it also drops with increased heat). Stronger currents are needed to produce more light, so a great deal of research and development has been centered on this issue. Most LEDs work best on direct current (dc), although many now are packaged with a driver that allows them to function with alternating current (ac).

There are two main ways to produce bright white light with LEDs. One is to use individual LEDs that emit three colors—red, green, and blue—and then to mix the light. The other method, which is much more common, is to use a phosphor coating to convert light from a blue or ultraviolet (UV) LED to white light, similar to the way fluorescents work. Each LED unit emits light in a narrow range of wavelengths, which is what makes them so efficient for signs and other colored applications. Phosphor-based LEDs are slightly less efficient as a result of the coating, but they are constantly evolving.

Normally, the color of the plastic lens is the same as the light actually produced by the LED, but this isn't always the case. Infrared (IR) LEDs often have purple cases, whereas blue LEDs often have clear cases. LEDs also can be made with two (bicolor) or more diodes in a single unit, which allows for switching and mixing between colors. Multiple LEDs can be tightly arranged in the same direction to produce beams.

LEDs can be made very small, for tiny indicator lights, or larger, for lighting (Figure 5-4). These can be driven with an ampere or more of current. This circuitry produces a considerable amount of heat (not unlike a laptop), which must be dissipated effectively or the LED will burn out quickly. Large LED "light bulbs," such as PAR (parabolic aluminized reflector) lights, have sizable heat sinks built in to get rid of this heat, and they do get quite warm to the touch

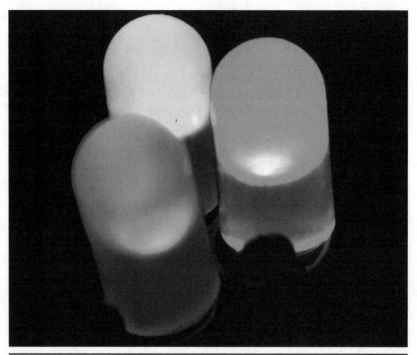

FIGURE 5-4   Lit LEDs. *(Wikimedia Commons)*

during operation (although the overall heat output is significantly less than that of an incandescent).

Environmental Benefits of LEDs

LEDs offer a number of promising environmental benefits, and they are often viewed as the future of green lighting. According to the U.S. Department of Energy (DOE), LEDs eventually could reduce by 50 percent the amount of energy we use for lighting globally and reduce by 10 percent the total amount of electricity we use. LEDs could reduce projected 2025 global carbon emissions by 300 megatons per year while creating new jobs.

To break this down, on average, generating a single kilowatt-hour of electricity produces 1.34 pounds (610 grams) of CO_2 emissions. If the average light bulb is on for 10 hours a day, a single 40-watt incandescent bulb will generate 196 pounds (89 kilograms) of CO_2 a year. However, a 13-watt LED equivalent will only produce

63 pounds (29 kilograms) of CO_2 over the same time. A building's carbon footprint from lighting therefore can be slashed by 68 percent by swapping out all incandescents for LEDs. It's also true that the long life of LEDs should mean that fewer resources will be needed to produce and maintain lighting equipment. And the fact that LEDs lack mercury is a clear advantage.

It is perhaps not surprising, then, that governments and private companies around the world have invested millions of dollars in research on LEDs.

Concerns with LEDs

Although there is a great deal of excitement about LEDs, there is also a considerable amount of hesitation in part because the technology is relatively new for lighting and because advancements are happening so fast and people are reluctant to buy something that they feel could become out of date quickly. "A lot of people ask about LEDs," Lou Nowikas of Hearst Corporation explained. "They're not there yet, but we're looking at them. There are new developments announced practically every week. We're looking at putting in some little ones in the elevators and are considering LED theater lights," he said.

Cost

Let's get the big one out of the way first. It's certainly true that LEDs cost more than conventional lighting right now—at least when we're talking about upfront prices. However, it's essential to look at the total cost of ownership, including energy and maintenance costs, and in that case, it's not the price of the bulb that is the biggest piece of the pie. In their lifetimes, LEDs save a chunk of change over incandescent and halogen lighting, and they are starting to get close to competing with fluorescents.

This is not to say that LEDs are for everyone and every application right now. David Bergman, the Leadership in Energy and Environmental Design (LEED) Accredited Professional (AP) architect and designer of the green lighting line Fire & Water, said that he doesn't think LEDs make that much sense for typical home owners just yet, although he has recently started to design a few fixtures for them. Bergman's cautious-yet-optimistic approach to the technology is common in the industry. "Right now, CFLs are a better buy when you're talking about general lighting for the home," Michael Smith, vice president of the Energy Solutions Group for lighting controls-maker Lutron, said in a phone interview.

LEDs are still relatively expensive to buy because of the high-tech components and relatively rare materials used, such as sapphire or indium. There are also basic economies of scale: Not that many lighting-quality LEDs have been built compared with other, more established products. But prices have been dropping rapidly. Just a couple of years ago, a 60-watt equivalent replacement bulb cost around $100. Today, you can order a ZetaLux LED bulb from Advanced Lumonics' EarthLED for $40, less than half as much! The ZetaLux uses only seven watts for the light output of a 50- to 60-watt incandescent bulb, so it only costs around $2 a year to run. According to EarthLED, in 10 years, the total cost of using that ZetaLux bulb would be $59.99, whereas it would cost $206.20 for the 30 incandescents it replaced (Figure 5-5). (That's $30 for the 30 bulbs and $175.20 in electricity costs, assuming 10 cents per kilowatt-hour, although, in reality, electricity rates are likely to go up.)

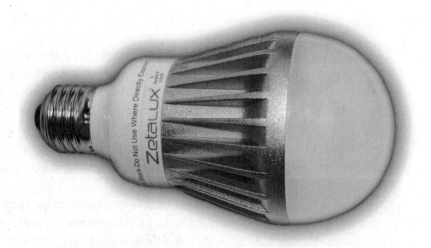

Figure 5-5 The ZetaLux LED "retrofit" bulb works in regular sockets but uses only 7 watts to approximate a 50- to 60-watt incandescent bulb. (*EarthLED.com/Advanced Lumonics*)

Sales of such retrofit LED bulbs are rising because the economics do work out favorably in the long term. However, it's impossible to ignore that the initial sticker shock is still a deterrent, particularly in a recession, when budgets are tight. Still, most experts expect LED costs to continue plummeting and quality to rise. It's also true that integrated LED lighting systems can work better than retrofit screw-in bulbs, so LEDs can be more attractive for new construction.

In 2007, the global market for LED lighting reached $337 million, up from $205 million in 2006, according to a report by Strategies Unlimited. The majority of uses in 2007 were for architectural lighting, signs, and accent lighting. Strategies Unlimited predicts that the worldwide market for LEDs will reach $1.65 billion by 2012. The DOE and the Optoelectronics Industry Development Association have estimated that by the year 2025, LEDs will be the preferred method of lighting in homes and offices.

Alternative Pricing Plans

One particularly exciting option for switching to the technology is LED Saving Solutions from Devon, Pennsylvania–based GREENandSAVE. The company offers commercial customers lease-to-own programs for LED lighting, as well as an innovative option that costs participants no money up front. GREENandSAVE technicians will come out to a facility, install LEDs, and set up energy monitoring. The company then gets paid according to a sharing agreement with the customer, with the money deducted from the customer's savings earned through lower utility bills.

So far, clients have come from a wide range of industries, including industry, retail, hospitality, health care, and education. Properties have ranged from several thousand to over a million square feet.

Dimming, Brightness, and Light Quality

Many lighting designers complain that LEDs aren't quite bright enough to really serve as general lighting. There is some truth to this, and most bulbs on the market don't really seem to match up to the level of the incandescents they are supposed to replace, at least at first glance. Part of the problem is that LEDs are most efficient at directing light in one direction. This is why they work so well for flashlights, which are now widely available and inexpensive. But the directed beam of a flashlight is not well suited for general lighting, in which you want some "wasted" ambient light. Most LED bulbs sold for retrofitting fixtures produce more directional light than people are used to. One possible workaround is to try LED reflector lights or spotlights, which take advantage of this property.

LEDs are getting better at color rendering, but they still aren't quite up to the color rendering index (CRI) of 100 produced by incandescents. The ZetaLux has a CRI of 75 in cool white and 80 in warm white, which is getting pretty good and which is competitive with most fluorescents. Some people still find LEDs on the cool side, however, and they complain that the light isn't that flattering. Further, the color temperature of LEDs can vary significantly—on the order of 500 K—according to different viewing angles, as well as quality control by the manufacturer.

Most LEDs are dimmable, although some on the market aren't compatible with standard dimmers. It's a good idea to ask before you buy. Again, this is an area where integrated LED systems, with their own dedicated controls, can provide more exact specifications.

Lifespan

There is often a tradeoff between LED lifespan and other variables, notably size and light output. By adding a larger heat sink, a LED likely will last longer, but this can make the whole package unwieldy, to the point where it wouldn't fit in many fixtures. Similarly, the brighter a LED unit is made, the more strain is put on it, often shortening its life. LEDs made to last for hundreds of thousands of hours simply aren't bright enough to use as lighting—at least not yet. Therefore, it's important to read the fine print when shopping. Some LEDs are marketed with an estimated lifespan, but with the caveat that they be operated only in exposed fixtures or with strong ventilation. If you put such a LED light in a recessed can, it probably will last a lot less long. This is so because high temperatures shorten the life of LEDs, although some models can be designed to be more tolerant. It's also a good idea to ask the manufacturer if the stated lifespan is for the unit as a whole or just for the core LED chip because this can make a big difference.

Although LEDs do last a very long time, like everything else, they eventually fail. Usually this is by gradually going dimmer, which can give the user a chance to react, versus the immediate burnouts of incandescents. With white LEDs, the phosphors do degrade eventually, causing light quality to decay, although this generally happens more slowly than degradation of gas-discharge lighting.

Light Pollution

If LEDs are used as significant outdoor light sources—which is not currently common—they should be shielded to reduce light pollution. This is so because LEDs emit more blue light than most conventional outdoor lights, and this contributes more to the glow in the sky that can obscure stars and disturb wildlife and other people. If LEDs will be used outdoors, they should be shielded according to the guidelines of the International Dark Sky Association.

Uses for LEDs

LEDs are small, light, durable, and versatile, and their applications are really limited only by our imaginations. Here are some of the more common uses today.

Indicator Lights

Since LEDs are small, efficient, and last a long time, they make ideal indicator lights, especially because they don't have to be bright but can be different colors. This was the first practical application of LEDs, and they are more common than ever, showing up in thousands of electronic and consumer products, as well as commercial equipment.

Backlighting of Displays

LEDs are a core technology lighting up televisions, computers, cell phones, watches, and many other devices. This includes Apple's groundbreaking iPad, which uses white LEDs in edge-lit configuration, allowing the tablet to be only half an inch thick (Figure 5-6).

FIGURE 5-6   The screens of Apple iPhones and iPads are lit with LEDs, like many other electronics devices. The iPhone 4 also has an LED camera flash. *(Photo by Brian Clark Howard)*

Auto and Marine Lights

Since LEDs are extremely rugged, efficient, and long lasting, they work great as secondary lights on vehicles. They aren't used commonly as headlights yet because those need to be extremely bright.

Signs and Traffic Signals

Since LEDs are small, produce directional light, can be colored, and can be switched or strobed rapidly, they work great for displays.

"We replace 150-watt incandescents or halogens with five- and eight-watt LEDs. That's about as green as you can get," Roy Burton, the CEO of Dialight, explained via telephone. Dialight is a British company that does a lot of business in the United States, and Burton is based in New Jersey. Dialight has been supplying LED traffic lights for 10 years and is the world's biggest provider. According to Burton, about 70 percent of U.S. traffic lights and about 10 to 15 percent of Europe's traffic lights are now LED lit (Figure 5-7).

Asked about recent news reports that LED traffic lights have caused accidents in the winter because they don't get hot and therefore don't melt snow, Burton responded: "If you look at the accidents caused by failed traffic lights, compared to LEDs, which have excel-

Figure 5-7 LED traffic lights are rapidly taking over as the dominant technology. *(Photo by Brian Clark Howard)*

lent reliability, I think you'll find on balance that LEDs greatly decrease the number of accidents." Burton acknowledged that the snow issue is something his company is looking at, although he said he doesn't know of a solution that doesn't take more energy (he said all coverings tried to date have made the problem worse). "It's rare that we hear about it, but if you can't see the light, the sensible thing to do is to stop," said Burton. "On the good side, LEDs save a huge amount of energy. And we guarantee all our products for at least five years."

Street Lighting

LED street lighting is showing impressive advancements and is gradually gaining ground, although it is still relatively expensive. In March 2010, University of Pittsburgh researchers concluded the first cradle-to-grave assessment of LED streetlights. They determined that the lamps "strike the best balance between brightness, affordability, and energy and environmental conservation when their life span—from production to disposal—is considered."

One example is Philips Lumec's RoadStar, which won the 2009 NGL (Next Generation Luminaires) competition put on by the DOE, the Illuminating Engineering Society of North America, and the International Association of Lighting Designers (Figure 5-8). The RoadStar is designed to replace conventional high-intensity discharge (HID) lights and is up to 50 percent more efficient. With Philips' LifeLED technology, it is said to last more than 70,000 hours, comes with a five-year warranty, and can have twice the pole spacing of other systems. RoadStar is Dark Sky–compliant, can function at up to 122°F (50°C), and is said to produce high-quality light with low glare.

Industrial and Structural Lighting

Since LEDs stand up well to extreme conditions, they are the best option for many intensive settings, such as on offshore oil wells, on mining equipment, and in petrochemical plants (ironic, perhaps, that these are all fossil fuel-based businesses?). "In areas where there is a lot of vibration, or where they are turned off a lot, like in a railway car, you wouldn't use a CFL," said Burton. "You can bash LEDs around and they don't fail, whereas conventional lights can break and that's it," he added. Dialight sells a lot of LED lights for these applications.

FIGURE 5-8   Philips Lumec's RoadStar may represent the next generation of efficient street lighting. *(Philips Lumec)*

The company also sells LED strobe lights for cell towers because they are rugged, save energy, and last a long time. "It's a safety issue, if you have to climb 500 feet to change a light bulb," said Burton. "Ours are guaranteed for five years, and we expect a 10-year life and 20,000 hours or more. They use 60 percent less energy than high-pressure sodium or metal halide lights and usually pay for themselves within a year."

Stage and Strobe Lights

The rapid strobing ability of LEDs also makes them well suited for stage and party lights. LEDs can be finely controlled, and they are more reliable and predictable than other technologies.

Track Lighting

LED track lights only recently entered the market. But since LEDs are small and produce directional light, they would seem to be a good fit for this use (Figure 5-9). In a recent visit to the lighting district in New York City, we saw several attractive examples of LED track lights on display. At one store, LED lights started at $115 per light, whereas halogen track lights started at $27 each. At another lighting store down the street, LED track lights could be picked up for $50 each.

Figure 5-9   LED track lighting is a relatively new product, but it offers a number of advantages. *(Photo by Brian Clark Howard)*

Accent Lighting

LEDs work very well for accent lighting because they don't have to be extremely bright, and they are small and require little maintenance. They can be readily operated on low voltage. "LEDs function so much better at 12 volts, so it would even make sense to have a 12-volt lighting source at a house and then 110 volts for the rest of it," Robert McNeill, the North American account manager of ATG Electronics, explained in a phone interview. "That 12 volts can be captured by solar panels, stored in batteries, and then used for lighting," he said.

McNeill knows something about this because he spent seven years living off the grid in a cabin in a California canyon. "I used LEDs if I could get them," McNeill said. Mostly he used 12-volt incandescents powered by solar panels, as was his satellite TV and Internet.

Holiday Lights

In just the last few years, LED strand holiday lights have burst into the mainstream (Figure 5-10). They are now widely available nearly everywhere Christmas or Hanukkah lights are sold, and they only cost a few dollars more than traditional incandescent kits. Yet they last a really long time and have a reduced fire risk because they don't get nearly as hot as incandescents. A traditional 26-bulb string of holiday lights uses 125 watts and lasts only 1,000 hours, whereas the same size strand in LEDs uses 2.3 watts and lasts 20,000 hours. LED holiday lights are, on average, 90 percent more efficient than conventional brands, meaning that they cost only a few pennies to power for a season. They'll save the average family about $100 per season, according to the New York State Energy Research and Development Authority (NYSERDA).

Figure 5-10 LED holiday lighting brings seasonal cheer without using lumps of coal. *(1000bulbs.com)*

LED holiday lights come in a range of colors from red to blue, white, and multicolored, as well as styles from icicles to balls and cones. There are also solar-power options (more on this in Chapter 9). LED holiday lights start at under $10, but if you choose them, you'll be in good company: This is what adorns the giant Christmas tree in Rockefeller Center in the heart of New York City, as well as the official trees of more and more cities across the land.

Night Lights

LEDs also work very well as night lights because they last so long and are so inexpensive to run. These can be picked up at drug or hardware stores for just a few dollars, and you never have to worry about replacing burned-out bulbs again. One example is the multi-directional LED night light from Satco, which lasts an estimated 100,000 hours. The unit has a rotatable head so that you can direct the light where you want it, and it has a built-in photocell so that it turns on automatically when it gets dark and off again when it's light. The annual operating cost is estimated at less than $1.

Flashlights and Desklights

LEDs are already becoming the dominant technology in areas where compact size and efficiency are important and where light is needed in only one direction (Figure 5-11). Prices have fallen so much that such lights are becoming ubiquitous.

Fluorescent Tube Replacers

ATG Electronics has focused most of its research and development efforts on LED tube lights, especially a model to replace T8 fluorescents in commercial settings (Figure 5-12). "It's not price competitive with fluorescents yet, but it doesn't contribute to the 50,000 tons of mercury that get disposed of every year. In that regard it is a very environmentally friendly product," Robert McNeill, ATG's North American account manager, explained via telephone. The LED tubes cost around $40, whereas fluorescent tubes cost $3 to $4.

An earlier generation of ATG's lights contains 60 LEDs in each tube, although a newer product has 300 LEDs in each four-foot tube.

FIGURE 5-11   Efficient LED desklights have become widely available, as well as inexpensive. *(Photo by Brian Clark Howard)*

"This has a milky white lens and smaller LEDs, which give more even light, and solid lighting across, with no shadows in between [unlike the earlier design]," said McNeill. "For someone used to a fluorescent bulb, shadows can be a bit distracting," he admitted.

Retrofit Bulbs

New products bring the efficiency and other benefits of LEDs to existing light fixtures to replace incandescents or CFLs. These are the

FIGURE 5-12   A look inside an LED bulb, showing the actual LED chips. (*EarthLED*)

LEDs that some critics say aren't quite ready for prime time, although it is undeniable that they will pay for themselves in the course of their lifetime. To some people, that's enough, and to others, it's important to buy LEDs to reduce our impact on the environment today and to support an emerging technology that offers so much hope for the future. Those who are truly committed to greening their lighting may want to consider trying a few of these LED bulbs (Figure 5-13).

Figure 5-13 Using only 12 watts, Philips' high-tech EnduraLED is designed to replace 60 W incandescents in standard fixtures, for an energy savings of 80% (plus a lifespan that's 25 times longer). *(Philips)*

There is an increasing array of choices available. One exciting area is MR16 LEDs, which are already nearly cost competitive with halogen MR16s and which are used in track and other downlighting. ATG's MR16s are made with three LEDs each and are powered by 3.5 watts, compared with 20 or 50 watts for halogens. And they release a lot less heat and last a lot longer. Attractive candelabra LED bulbs are also available, again providing longer life and better efficiency than incandescents (Figure 5-14), although they haven't quite reached the brightness.

Edison screw-base LEDs continue to improve at a rapid pace. The $40 ZetaLux by Advanced Lumonics's EarthLED, for example, has a pretty respectable CRI of 75 in cool white and 80 in warm white. It has a power factor greater than 0.85, which puts it on par with high-quality CFLs and reduces stress on the grid (refer to Chapter 4 for more on power factor). The ZetaLux is also the first Underwriters Laboratories (UL)–listed LED retrofit for general household

FIGURE 5-14   LED accent lights can be great replacements for candles, and their flicker patterns are so natural that many don't notice they aren't "real" flames. *(Photo by Brian Clark Howard)*

lighting. The LED chip inside is made by Cree, a North Carolina–based company that has earned praise from President Obama and Vice President Biden for creating new green jobs.

EarthLED warranties its bulbs for two years, including the EvoLux line, which matches the light output of a 100-watt incandescent with a mere 13 watts. These are also available in two color temperatures, they last an estimated 50,000 hours, and they cost only $6 a year to operate (although they cost $80 to $100 to buy). Advanced Lumonics also offers a complete line of LEDs that work with standard dimmer switches, exhibiting a full range of dimming. These are called *Lumiselect*, and they come in several PAR and globe sizes.

Other good sources of LED retrofit bulbs include Superbright leds.com and Pharox. One thing you'll notice about these bulbs is that they are made with a fair amount of metal, usually flaring out from the screw base, either smooth or with ridges. This is the heat sink, which dissipates heat away from the LED's circuitry to avoid damage to the core components.

At the time of this writing, press releases from the world's biggest manufacturers have been heralding a new generation of mass-produced and widely available retrofit LED bulbs. GE, Philips and Sylvania say they are all coming out with high-quality LEDs to replace standard incandescents, and at more competitive price points (see Figures 5-13 and 5-15). Home Depot has announced plans to sell a 40-watt equivalent for $20 through its EcoSmart label.

Grow Lights

LEDs are also gaining increasing attention from plant cultivators, who are interested in the energy savings, as well as reduced heat, because high temperatures can damage sensitive vegetation. Therefore, LEDs are being designed and tested to provide optimal growth spectrums.

FIGURE 5-15   Osram Sylvania is also producing a 12 W LED bulb for general use, which the company says can replace 60 W incandescents. It will give off 810 lumens, with a color temperature of 2,700 K. *(Osram Sylvania)*

Liquid-Cooled LEDs

At the massive Consumer Electronics Show in 2010, a Taiwan-based company demonstrated liquid-cooled LEDs, marketed as Liquileds. It's an innovative idea, although some observers are skeptical. Jaymi Heimbuch, who writes about technology for Treehugger.com, wrote, "Sounds great, but if one of these things breaks ... what a mess. And these aren't really any different from the Hydralux-4 liquid-cooled LED being distributed by EternalLEDs that got such bad press not so long ago. Next generation of LEDs? Maybe, but probably not."

"There's all sorts of ways to cool LEDs. Earlier computers were liquid cooled," explained Roy Burton of Dialight. "Most manufacturers don't use liquid cooling, although we use semiliquid heat packs for some of our products. In a strobe you're dealing with a sealed environment, because you don't want water to get in there, but the heat is sealed in as well, so we use heat pipes to take that out. Some of these products are custom designed," he said.

Summary

LEDs offer tremendous promise for the near future, and they are increasingly available today. They already make smart sense for a lot of applications, from accent and holiday lights to flashlights, desk lamps, traffic signals, decorative displays, and more. Very soon we will be seeing more LED streetlights and spotlights. Within a few years, they probably will become the dominant technology for general lighting—but forward-thinking and progressive individuals need not wait for the rest of the crowd to catch up (Figure 5-16). Alternative pricing plans, such as LED Saving Solutions, make it even easier to get started, with bulbs installed for free and the company getting paid only through the savings you make.

Although LED retrofits have a high initial price today, they will pay for themselves in a few years. Think of all the carbon emissions you'll reduce and the hassle you'll avoid from having to change light bulbs. People don't think twice about lining up outside the Apple store and paying top dollar for the latest gizmo. If we could channel some of that hip, first-adopter energy to LED lighting, we could move more units faster and bring down the prices for everyone.

FIGURE 5-16   Kim Lancaster's green show home in Rhode Island features recessed LED lights to gorgeous effect. *(Ashley Daigneault/ Caster Communications)*

CHAPTER 6

Fixtures and Controls

Lexmark recently received Gold Leadership in Energy and Environmental Design (LEED)–certified status for its ink cartridge recycling plant in Juarez, Mexico. The company installed motion sensors and a programmable control panel in the manufacturing and warehouse areas to switch off lighting during off-shifts. Interior lighting levels were reduced, but the actual perception of the light is better owing to better color rendition of new high-bay fluorescent fixtures. As a result of the changes, Lexmark slashed lighting energy use at the facility by 20 percent indoors and 60 percent outdoors.

As we learned in Chapter 2, lighting fixtures must hold and position the lamp, as well as attach it to the energy supply. Although this task sounds simple and straightforward, there are tens of thousands of different fixture designs, in practically every conceivable shape, size, and color. Fixtures make a big impact when it comes to light quality and usefulness, as well as energy efficiency (and, of course, design aesthetics) (Figure 6-1).

To get more precise, the *fixture efficiency* is the ratio of light emitted from a fixture versus the light emitted by just the lamp (bulb)(s) contained in the fixture and is expressed as a percentage. There are a number of factors that affect this ratio, such as the color and types of shades and other components.

FIGURE 6-1   A halogen wall fixture is easily dimmable. *(Photo by Brian Clark Howard)*

Some additional key points:

- *Unclean light fixtures can increase energy consumption by 25 percent.* Thus, even if you install the best energy-saving bulbs, if you don't clean the fixtures regularly, you won't enjoy all the benefits.
- *Outdated or old fixtures can reduce the efficiency of your lighting significantly.* Perhaps not surprisingly, technology has improved. In most cases, it is possible to request a fixture's efficiency rating from retail staff or a manufacturer. Thus, if you are buying new or upgrading, look for the fixtures with the highest ratings. Similarly, it is often a bit more efficient to use dedicated fluorescent fixtures with compact fluorescent lamps (CFLs) if you can, versus regular old incandescent fixtures.

Types of Fixtures

When selecting a fixture, always choose one with a rating that is appropriate to where you will be using it. The most well-known ratings are maintained by Underwriters Laboratories (UL), an independent testing organization approved by the Occupational Safety and Health Administration (OSHA). UL was established in 1894 in Northbrook, Illinois to develop standards and test procedures for various products and materials. UL allows manufacturers to display its seal as long as they remain compliant with its stated standards.

Moisture Designations

UL and other similar organizations typically specify the following categories when it comes to moisture:

- *Dry rating.* A fixture with this rating should only be used in areas not exposed to moisture. Any fixture that is not explicitly listed for wet or damp applications should be considered a dry fixture.
- *Damp rating.* A fixture with this rating may be used in sheltered outdoor areas that are protected from direct contact with rain, snow, or other moisture.
- *Wet rating.* A fixture with this rating is suitable for outdoor locations that receive direct contact with moisture (such as rain, fog, or ocean spray), as well as in baths and showers.

Energy Star Certified

Most Energy Star–qualified fixtures come with pin-based CFLs, which make it more difficult for people to swap them out for less efficient incandescent bulbs after initial use. However, note that some Energy Star outdoor fixtures will accept an incandescent bulb because they save energy through a motion sensor and/or a photocell.

Energy Star–qualified fixtures are readily available in the marketplace and come in hundreds of popular styles, including floor, table, and desk lamps, as well as in hard-wired styles for ceilings, walls, bathroom vanities, and outdoors. Many support dimming, motion sensors, and automatic daylight shutoff. According to Energy Star, if every home in America replaced the five light fixtures it uses most with registered models, it would prevent greenhouse gases equivalent to the emissions of nearly 10 million cars.

Energy Star–qualified fixtures have these characteristics:

- They use one-quarter the energy of traditional lighting fixtures.
- They support bulbs that must last at least 10,000 hours (about seven years of regular use), which is much more than the typical 1,000 hours for incandescents.
- They must distribute light more efficiently and evenly than standard fixtures.
- They carry a two-year warranty—double the industry standard.

The following are the major different types of fixtures. In each category, individual models can range from highly decorative classic styles to ultramodern concepts and bare-bones utility designs. Note that many fixtures will work well with fluorescents as well as incandescents and that there are light-emitting diode (LED) and halogen examples for every category.

Floor Lamps

Floor lamps rest on the floor, of course. A *torchiere* is a lamp with a shallow, bowl-shaped light fixture mounted on a pole. Torchieres direct light upward and are usually free-standing floor lamps, but they also can be attached to a wall.

Table Lamps

As the name states, these are designed to rest on tables or other surfaces.

Sconces

Descended from the candlestick holders of yore, sconces attach to a wall, mirror, picture frame, or other vertical surface, and they hold the light source with a bracket. They are often used to create a warm, inviting look and are popular in living areas and hallways (Figure 6-2). One note for public building managers is that sconces should be compliant with the Americans with Disabilities Act, which mandates in Section 4.4 that "objects projecting from walls with their leading edges between 27 and 80 inches above the finished floor shall protrude no more than four inches into walks, halls, corridors, passageways, or aisles."

Figure 6-2   Wall sconces come in many shapes and sizes, and they add interest to rooms.

Vanity Lights

A new vanity fixture really can transform the look of a bathroom. A *vanity* is a counter that holds a washbasin, and there is often a fixture placed above it, anchored in the wall or a mirror. Instead of using one fixture directly over the mirror, which can create glare and harsh shadows, it's best to set two fixtures 35 to 40 inches apart and at head level (more on this in Chapter 8).

Pendant Lights

A *pendant* is a fixture that is suspended from the ceiling using a chain, rod, or "aircraft cable" (tough wire rope). The light source is usually placed inside a bowl or drum. Pendants are common in foyers and kitchens, especially over eating areas. CFLs and LEDs typically work well in them (Figure 6-3).

Figure 6-3 This whimsical Chalkboard Pendant by St. Louis–based artist John Beck is made with 95 percent recycled steel (more on green materials in Chapter 7). *(Photo by Gloria Dawson)*

Chandeliers

Perhaps more well known to the general public is the *chandelier*, which is also suspended from a ceiling. Chandeliers position the light sources on arms, which may leave the bulbs exposed, or cover them with a shade or globe. Chandeliers are popular for dining areas and entrance halls. There are many styles for every type of lighting technology.

Ceiling Fan Lights

These lights are built into fan units. Look for Energy Star–certified combos, which are about 50 percent more efficient than conventional models and will save you $15 to $20 per year on your electricity bill. Energy Star–qualified ceiling fan lights are available from more than 30 manufacturers.

Track and Cable Lights

Popular in urban lofts and practically synonymous to some people with 1980s decor, *track lighting* is a strip of metal that attaches to the ceiling and provides power and support to fixtures that hang from it and can be slid to any position along the length. *Rail lighting* is just another name for this scheme, although some designers prefer the term to describe the type of strip used. A *monorail*, for example, has a single track, whereas pairs of tracks are also common. *Cable lighting* is similar, but instead of attaching to the ceiling, cables are run from one wall to another, and fixtures are hung from them. Cable lighting is often used for rooms where you can't mount anything to the ceiling.

Track lights have come a long way since the 1980s, and they offer considerable flexibility. Fixtures can be slid along the track to highlight artwork or other features as needed, and they can be angled in many directions. They work well as general or task lighting. Many track systems now use low-voltage bulbs (more on this later), which means that a transformer must be present to step down the line voltage. Low-voltage track lights are safer to touch, and they have more flexibility of design—many are on curved tracks and sport highly decorative fixtures—but each circuit can support only a limited number of lights.

These days, most track lighting systems use halogen bulbs, which can produce a lot of light with a small bulb (Figure 6-4). Small CFL reflector bulbs are now available as replacements, however, offering greater energy efficiency and producing less heat. LED track lighting is a relatively new product on the market, but a lot of designers are excited about its potential.

Flush-Mount Fixtures

These fixtures install directly against the ceiling without a stem. (Sometimes versions with short stems are called *semiflush*.) The pan, or body, of the fixture attaches to the electrical junction box, often

FIGURE 6-4   Halogen track lights. *(Photo by Brian Clark Howard)*

with a hanger bar mounted to the box with screws. A threaded rod then screws into the hole in the center of the bar, which aligns with a hole in the bottom of the glass and is secured with a decorative knob called a *finial* (this isn't always visible, depending on the design). The glass also can be secured to the fixture with screws on the outside, or it may have a twist-to-lock feature. As with many other fixtures, flush mounts can be decorative or utility.

Suspended Downlights

These fixtures hang from the ceiling and direct most of the light downward. A good green choice is to use LEDs or CFLs to replace incandescent A-lamps.

Recessed Downlights

These fixtures also direct light downward but are set into the ceiling (or false ceiling), often in what are called *cans* or *high hats*. Although originally intended for task lighting, recessed downlights have become very popular and are now used widely for general ambient lighting in kitchens, hallways, bathrooms, and many other areas of homes and businesses.

As stated previously, it is preferable to use a reflector CFL over a regular spiral both to decrease the chance of damaging the bulb with heat buildup and to more efficiently direct the light downward, where it is needed. Putting a standard bulb in a can will waste a good portion of the light.

Under-Cabinet Lighting

These fixtures direct light down toward a work surface. Common examples include linear fluorescents, which should fill at least two-thirds of the cabinet width, or strip LED lights, which are becoming increasingly popular.

Architectural Luminaires

Architectural luminaires mount to a ceiling or wall, but the lamps are shielded behind a board. They produce flattering light that is great for living rooms, bedrooms, kitchens, and bathrooms, as well as commercial spaces. Fluorescent tubes are commonly used with them. The shielding boards can be made from one-inch lumber stock, plywood, metal, or drywall, which can be painted, stained, or covered with fabric. The inside surface should be painted with semigloss white paint to reflect the light. Luminaires can be left open above and below the shielding board, or they can be covered to more completely conceal the bulb (with baffles, louvers, and diffusers). Luminaires are relatively easy to build and fasten.

Types of architectural luminaries include the following:

- *Soffits.* These direct light downward. They can be used as direct lighting over a table or counter or as general lighting in rooms with low ceilings.

- *Valances*. These direct light downward and upward. They work well for ambient lighting and are seen often in concert halls and other large spaces.
- *Coves*. These direct light upward. They work well with high ceilings and above kitchen cabinets.

High- and Low-Bay Lights

High-bay fixtures are designed to illuminate large spaces and are placed at 12 feet or higher above the surface (Figure 6-5). Related to these are *low-bay fixtures*, which are also designed to illuminate large spaces but are installed at heights below 12 feet. These are most common in warehouses and industrial settings.

Outdoor Lighting

Done well, outdoor lighting can enhance safety and aesthetics with minimal disruption. Remember to use qualified outdoor fixtures and components, including bulbs that are rated for outdoors. If you live in a cold climate, make sure to use fluorescents that are designed for your temperature range. Standard CFLs do not work well below

FIGURE 6-5 High-bay fixtures are designed to light large spaces.
(Lithonia Lighting)

40°F (4ºC). Controls are particularly important when it comes to outdoor lighting, and we'll get to them shortly.

Some key terms:

- *Floodlights.* These are designed to spread light uniformly over a large area. They are often mounted high on a wall, such as above a garage, but they can be on a pole or on the ground.
- *Spotlights.* These focus a bright light on a single subject, just like on the stage. They can be mounted on a wall, a pole, or the ground.
- *Postlights.* These are lights mounted on posts.
- *Pathlights.* These are designed to direct light low to the ground, with the illumination ideally spreading horizontally over the path surface.
- *Lanterns.* Enclosed by glass, these fixtures resemble old-fashioned gas or candle lanterns, and they can be mounted against a wall or on a pole.
- *International Dark Sky Association certification.* This Fixture Seal of Approval is a third-party certification for luminaires that minimize glare and reduce excess light pollution.

Low-Voltage Lighting

One option that is becoming increasingly popular and that offers some definite green benefits is *low-voltage lighting*. To the user, the result is just as effective as standard *line-voltage lighting*, although the difference is in the details. Low-voltage lighting can work well for indoor or outdoor applications and for task, accent, or general lighting. Many better lighting stores carry low-voltage fixtures and components.

Low-voltage outdoor lighting got a colorful start back in the early 1950s. A California contractor named Bill Locklin was asked by a client to put out some yard lights. So Locklin cobbled together some fixtures out of old juice cans, tractor headlamps, and mayonnaise jars powered by 12 volts from car batteries. Locklin's client was publishing mogul Walter Annenberg, who was entertaining President Dwight D. Eisenhower and his wife. The first couple were so impressed by the unique lighting that they ordered their own system. Locklin's

Figure 6-6   Low-voltage landscape lighting by California-based Nightscaping offers many benefits, as well as beauty and utility. *(Nightscaping)*

company, Nightscaping, was off to a bright start, and it is still going strong (www.nightscaping.com) (Figure 6-6).

By definition, low-voltage lighting operates at 30 volts or less. It offers a number of benefits, including greater safety, ease of installation (do-it-yourself work is popular), greater flexibility of designs, increased energy efficiency, and enhanced light output. Low-voltage lighting is the natural choice for many settings that are off the grid. Otherwise, what you need, in addition to appropriate fixtures, is a transformer to step down the voltage from the line (say, 120 or 240 volts) to 12 or 24 volts. The transformer can be integral (built into the fixture) or remote.

When set up properly, low-voltage lighting produces 2½ times as much light as line-voltage incandescent lamps. In other words, a 50-watt low-voltage lamp generates as much light as a 125-watt line-voltage lamp. This means that each lamp saves up to $7.50 per 1,000 hours of use, assuming an energy price of 10 cents per kilowatt-hour.

Low-voltage systems offer exceptional flexibility, making it easier to design curved and complex track systems and landscaping lighting plans. When working with a remote transformer, low-voltage fixtures can be made very small, offering more aesthetic choices.

The technology can readily be dimmed, and LEDs work particularly well with it. And perhaps best of all, low-voltage lighting extends the life of bulbs.

Low-voltage lighting (Figure 6-7) can be particularly attractive for commercial settings in part because it offers such fine control of lighting. And it can reduce waste by extending bulb and fixture life. High-quality remote transformers often carry warranties of 25 years.

Figure 6-7 Low-voltage fixtures on display in New York City's lighting district. *(Photo by Brian Clark Howard)*

Lighting Controls

It can be a pain to try to remember to turn off lights when you're not using them, and this can lead to arguments and finger pointing in the home or office. So why not give this repetitive, relatively thankless job over to a machine? Effectively using lighting controls should result in substantial energy savings, as well as a reduced environmental impact.

There is an ever-increasing range of products for controlling lighting down to the finest details. Large facility managers can install sophisticated systems that give centralized control over thousands of fixtures, resulting in savings of hundreds of thousands of dollars a year. Home owners can install a whole-house smart system, spending more than a thousand dollars, or they can spend much less for a panel in their bedroom that gives control over half a dozen main lights. For $100, you can get a wireless remote that allows you to turn house lights on and off from your car so that you never have to come home to a dark house. Or you can spend just a few dollars for a simple wall dimmer switch or timer.

The most common types of lighting controls include:

- Dimmers
- Motion and occupancy sensors
- Photosensors
- Timers

Let's take a closer look.

Dimmers

Dimmer controls reduce the brightness of light, which saves energy. Period. "Even if you never use the dimmer you installed, it saves about four percent of the energy for that light," explained Michael Smith, vice president of the Energy Solutions Group for Lutron, the world's leading lighting controls company. (This is because Lutron's dimmers automatically dial back the energy used, but without diminishing light quality, according to Smith.) "Our dimmers are solid-state devices with a processor that switches the lights on and off. A dimmer turns light on and off 120 times a second; the eye just can't see it. The more you dim, the more cycles the bulb is off. If you are dimming 50 percent, the lights are off 50 percent of the time. The more you dim, the more it's off, and the more energy you save," Smith explained via phone from Lutron's head office in Pennsylvania (Figure 6-8).

Lutron makes lighting controls for every type of light source, from incandescents to fluorescents, LEDs, and more. With all lights, the more you dim, the more energy you save, and the longer you extend the life of the bulb. According to Smith, the relationship between lower light and savings is nearly one to one, although it's a bit less efficient with incandescents. "A dimmer is one of the best energy-saving devices, and they also make your home more beautiful. You're not sacrificing; you're choosing the light level that's best for you," said Smith.

Off-the-shelf dimmers for incandescent fixtures cost only a few bucks. Fluorescents cannot be dimmed, unless they work with a dimming ballast, either built into the "bulb" or present as a separate unit, which may be part of a designated dimming switch product. As discussed in Chapter 4, dimmable CFLs continue to improve and

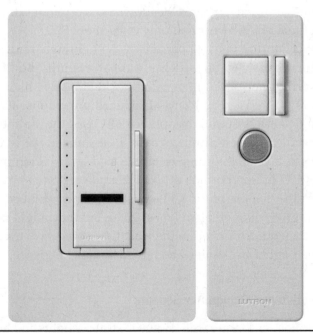

FIGURE 6-8 Dimmers have more style and flexibility than ever, and many work wirelessly. *(Lutron)*

are widely available. In addition, cold-cathode CFLs can be dimmed, as can some LED products.

A number of different dimmer designs have been used since the late 1800s, including saltwater dimmers, coil-rotation transformers, rheostats, thyristors, and various digital products. Joel Spira—founder of Lutron—invented the first solid-state dimmer in 1959, based in part on his experience working with radar in World War II. Detailed explanations of these technologies are beyond the scope of this book. However, consumers will notice that lighting stores stock quite a range of dimmer panels, which boast paddles, wheels, touch pads, buttons, and other convenient interfaces. Some even incorporate LED display lights. Installing a dimmer switch is pretty quick and easy (there likely will be only three wires to connect, and any product you buy should have instructions). If you are in doubt, seek a licensed electrician. If you do go do-it-yourself (DIY), make sure to disable the power to the circuit first!

Lutron offers a number of its most popular dimmer styles with Eco-dim, which boasts 15 percent energy savings over standard switches, even if you never turn it below full power. Eco-dim also allegedly extends the life of incandescent and halogen bulbs three times or more. Lutron's Meadowlark Eco-minder dimmer saves an estimated $30 a year and only costs about $25.

Unfortunately, 90 percent of the lighting circuits in the world still use basic on-off switches, estimates Smith. "But if everybody in North America put in one dimmer, it would be like taking 470,000 cars off the road in terms of reduced greenhouse gas emissions," said Smith [assuming people actually used the dimmers]. Currently installed Lutron dimmers save an estimated 9.2 billion kilowatt-hours a year, according to Smith, saving $1 billion in energy costs and 17 billion pounds of CO_2. "People just aren't aware," argued Smith. "People put in CFLs to save energy, and because they saw the green marketing and messaging, but they might not realize that if they put a dimmer on the wall, they could save even more, make their lights last longer, and have a better experience."

Motion and Occupancy Sensors

Motion sensors are well known in security lights and burglar alarms, but they are also becoming increasingly popular inside homes and in office buildings and other commercial settings because they are convenient and can result in substantial energy savings. Simply put, a motion sensor turns a light on after motion is detected and turns it back off a short while later. How does it detect motion? A sensor can use infrared (IR) scanning to look for body heat, ultrasonic pulses (similar to sonar), or microwaves (similar to police radar guns). An occupancy sensor is a motion sensor armed to shut off a light after the device no longer detects that someone is present.

Photosensors

Photosensors work great for outdoor lighting because they can switch it on automatically when it gets dark and turn it off when it gets light. This can result in considerable energy savings. Photosensors are popular for accent, path, street, commercial, and security lighting, as well as for night lights.

Timers

Timers, naturally, are used to turn on and off lights at specific times. Timers are rarely used alone for outdoor lighting because they would have to be reset often owing to seasonal variations in day length (unless you live near the equator). However, timers can be paired effectively with photosensors. For example, path and driveway lighting could be switched on in the evening by a photosensor and then switched off at a specified time later in the night by a timer, say, at 2 a.m.

Timers (Figure 6-9) are also often used for security reasons to make empty houses look occupied. They don't work so well for occupied spaces because people rarely stay in one spot for a set amount of time. Timers can be old-fashioned mechanical designs or newer electronic models.

Integrated Fixtures

Various sensors can be wired directly to fixtures, or they can be built in by manufacturers. An example of the latter is Occu-smart by New York–based Lamar Lighting. This system incorporates an ultrasonic motion sensor into a fluorescent wall- or ceiling-mounted luminaire

FIGURE 6-9   Timers provide security, convenience, and energy savings at the push of a button. *(Photo by Brian Clark Howard)*

that includes an electronic ballast. Occu-smart fixtures are designed for stairwells, storerooms, corridors, and other seldom-used areas. The units are always on standby, putting out a low level of light for safety and security. When the sensitive motion sensor picks up movement, the unit switches the light on to full brightness. After a predetermined, and adjustable, length of time, the light dims back to standby.

Wireless Sensors

Many sensors now offer the convenience of wireless operation. One example is the Radio Powr Savr from Lutron, which, according to the manufacturer, installs in minutes. The sensor communicates wirelessly with compatible dimmers and light controls and can be moved around for optimal coverage. It reportedly can "see" 676 square feet when mounted on a 12-foot ceiling.

Wireless sensors may use the ZigBee Protocol, a global wireless language that is being used for an increasing array of devices (similar to Bluetooth). Or they may use proprietary communications protocols.

Whole-House and Whole-Building Systems

Remember that different lights can have separate controls, even within a single room. This allows maximum flexibility and can facilitate greater energy savings. Many people are taking this a step further and are investing in whole-house or whole-building lighting control systems.

Kim Lancaster and Joe Hageman are happy with the Lutron HouseWorks whole-house system in their Rhode Island green dream home. The system is programmed with a number of one-touch presets, which can light up specific rooms or combinations of rooms or a path to the kids' bedrooms from the master bedroom. There is an "on/off" button for all lighting, as well as a "home" button that turns on the entryway and kitchen lights, and an "entertaining" setting, which sets a pleasant mood. The family seamlessly operates its LED holiday lights on a timed schedule, and they have motion sensors for outdoor fixtures, including a sensor embedded in the driveway. Everything is dimmable, and the family has the default brightness set at 85 percent, instead of 100 percent, to save energy.

The smart lighting is just one part of the centralized controls in Lancaster's 4,400-square-foot Green Life Smart Life home, which is managed in its entirety by an integrated Control4 system. This also governs the geothermal heating, ventilation, and air-conditioning (HVAC) system, as well as security, entertainment, and appliances. The resulting energy savings are substantial. After a device is switched off, sensors automatically cut the power going to it, heading off the so-called vampire effect of standby power drain. When rooms become unoccupied, sensors turn off certain devices. The refrigerator is automatically shut off each night from 1 to 4 a.m., although food is still kept cool. The whole-house audio system, Essentia from NuVo Technologies, is Energy Star–rated. The entire home system can be tuned to automatically reduce energy use further— for example, you can set it to disable certain functions when peak periods result in expensive electric rates. The system provides real-time energy-use information, making it possible for occupants to make adjustments on the fly and to see exactly what they are paying for.

Control4 operates seamlessly with Lutron's systems, as well as the other components of the smart home. If and when a smart meter is installed by the local utility, it will provide even greater flexibility. Lutron HomeWorks now can be installed as retrofits in older and historic homes, as well as in new construction. Many of these systems can be controlled via the Internet and cell phones, making them flexible to changes in your schedule. Walk out the door in the morning and forget to turn your bedroom light off? No problem! Do it via text from your phone in seconds.

"We get people very actively involved in energy management of their homes," Jay McLellan, president and CEO of Home Automation, Inc., recently explained at the Greener Gadgets 2010 conference in New York. "We work with upscale customers who want full-house audio and other conveniences, and we use low-power equipment. If you decrease the comfort of a home, you're going to get put out on the back porch with the milk bottles," he said. McLellan's New Orleans–based firm designs and installs a range of integrated energy and technology management tools.

On the commercial level, more managers are switching to smart buildings, which, according to Michael Smith, vice president of the Energy Solutions Group for Lutron, can result in energy savings of 40 to 70 percent. Many are setting their lights to autodim no higher than 90 percent, which saves money and is not noticeable, according to Smith. Motion and occupancy sensors pay for themselves quickly in commercial settings. Another technique is to engage in *demand response*, or *load shedding*, which means that once a manager is alerted by the system that electric rates have risen during a peak time, he or she can activate a control to, for example, dim lights or dial down a thermostat. Such real-time response can save serious cash.

A Boston-based startup, Digital Lumens, says that it can achieve energy savings of 90 percent over standard high-intensity discharge (HID) lights in warehouse and commercial settings by using networked LED light fixtures. The company installs a system of sensors and software that controls precisely how and when the lights are used.

Summary

The bottom line is that controls and sensors give you greater flexibility to design a lighting plan that better serves your needs while reducing your energy use. With a few clicks of a mouse or a touch of a button, you can:

- Lower light levels to save energy and increase bulb life.
- Set the best mood for the time and place.
- Increase functionality and security.
- Save time and hassle.

Now that you are aware of different green lighting technologies, you can put them together with fixtures and controls to make them even more efficient and productive. Read on for advanced lighting strategies and how to make everything even more sustainable.

CHAPTER 7

Beyond Energy to Green Lighting Materials and Processes

Now that you've swapped out those dusty old incandescents for something sexier and more efficient, what do you do with the old bulbs? A reader of *ReadyMade* magazine has a timely suggestion: Turn them into flower vases! With a tap of a hammer and a few snips of some pliers, you can amaze your friends with a funky, attractive piece of do-it-yourself (DIY) decor. And with the phaseout of incandescents looming on the horizon, what better symbol could there be than an old-fashioned light bulb turned upside down and made into something that nurtures plants (Figure 7-1)?

Another reflection on the dimming of incandescents into history is offered by London-based artist Tim Fishlock. His recent piece—*What Watt?*—is a unique spherical chandelier made from 1,243 incandescent light bulbs of various shapes and sizes that is lit with a single energy-efficient light. According to Fishlock, the piece "marks the passing of a design that has remained relatively unchanged since its invention 130 years ago."

While it's easy to wax poetic about fading incandescents and perhaps even to reimagine them in creative new ways, we are only just beginning to scratch the surface of sustainable lighting materials. There are as many possibilities as rainbows reflected from your grandmother's crystal chandelier.

According to New York City–based architect and green lighting designer David Bergman, "It only makes sense for people who are concerned about energy efficiency to also ask for green materials in

FIGURE 7-1   This soon may be the best use for an incandescent light bulb. *(Photo by Brian Clark Howard)*

lighting." For one thing, Bergman explained in a phone interview, "Manufacturers are going to become more aware of liabilities in using toxic materials."

To date, there hasn't been a lot of concern on the part of the industry when it comes to greening up the materials that make up lighting hardware, with the notable exception of the drive to decrease the mercury content of fluorescent bulbs. Beyond that, however, manufacturers have been focused mostly on improving the energy efficiency of their products. This is a critical goal because the lion's share of the impact from lighting comes from the energy use of the products themselves, not from making them or shipping them to market. Still, as industries and environmentalists are increasingly learning, we need to take a careful look at each step along the long, interconnected chains that form our modern society.

"Environmentally sustainable product design and sustainable business practices should be more fundamental in our practices," Bergman argues. He pointed out that current Leadership in Energy and Environmental Design (LEED) guidelines do not consider the recycled content or other possible green features of lighting fixtures, only how energy efficient they are. Bergman, who is a LEED AP- (Accredited Professional) certified architect, explained that he has asked the U.S. Green Building Council (USGBC) to consider expanding this position.

Here are a few questions to keep in mind as you search for ecological light fixtures:

- Is the fixture durable and long-lasting?
- Is it made from recycled, recyclable, and/or sustainable materials?
- Does it have nontoxic finishes and components?
- Can I use less materials with it?
- Will it complement the space I have in mind, and can I possibly reuse it in different spaces?
- Is the company that produced it engaged in sustainable business practices?
- Is it locally sourced?

A Visit to Fire & Water

With his own boutique line of lighting fixtures, Fire & Water, Bergman continues to push the envelope of sustainability. Bergman also teaches green design at the Parsons School of Design and is the cocreator of "Educating the Educators: A Crash Course in Ecodesign."

On a recent visit to Bergman's third-floor walk-up apartment in Manhattan's Lower East Side, he told us more about his design philosophy and showed us some of his unique lighting creations. "When I started doing lighting back in the early 1990s, you couldn't use CFLs [compact fluorescent lamps] because they were terrible," Bergman explained, as his little dog, Luna, peered out from under a chair. "So I did then what I now call my 'legacy lines,' which used halogens and incandescents. But I always wanted to do energy-efficient lighting."

Today, Bergman's pieces are made almost exclusively from sustainable or recycled materials. Several of Bergman's fixtures incorporate a biocomposite made of recycled newsprint and soy flour. It has a smooth finish, not unlike dark lacquered wood. Bergman also employs recycled glass and plastic, as well as sustainably harvested wood, plus the occasional found object. Metal is recovered scrap that's shaped and polished to look new.

Yet Bergman's fixtures are clean, contemporary, and stylish, and they certainly don't look like anything you'd imagine coming out of a junkyard. (The prices aren't either; at $400 to $2,000 retail, they are aimed at what Bergman calls a "medium-high-end clientele.") Hanging in Bergman's living room were two examples of his graceful, elegant Fibonacci series, which boast a touch of intellectual whimsy. Bergman explained that the curvature of the shade is designed to "give the light a way out." The shade is made with a thin wood veneer (sustainably harvested) mounted on a paper backing, which is a notable alternative to the plastic backings most conventional designers use. Either in the overhead or wall sconce version, the light from a Fibonacci fixture is warm, diffuse, and comforting; you wouldn't know that it's a bright CFL unless you peaked under the veneer skirt (Figure 7-2).

In fact, Bergman prefers to build fixtures that have dedicated CFL pin connections so that consumers won't be able to swap in incandescents down the road. "You can't really call something green lighting without a dedicated base," he said. Bergman also showed off his Flipster lamps, which—like most of his creations—come in floor, ceiling, wall, and table versions (Figure 7-3). Made of shades with 40 percent recycled content, Flipster fixtures have flaps that can be opened and closed seamlessly, allowing flexibility to bring more or less light into a room or to bounce it off opposing walls for a soft theatrical effect.

Bergman also demonstrated larger pieces from his Parallel Universe series, made from polycarbonate and chips of colored recycled glass that are arranged by hand in unique patterns. Fitted with a dimmable CFL ballast, the effect is impressive—not quite as smooth or to the full range of a traditional dimming incandescent but a fairly

FIGURE 7-2   David Bergman's Fibonacci series is designed to channel light with shades made of sustainably harvested wood veneer and paper. *(Fire & Water Lighting/David Bergman)*

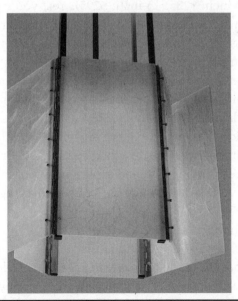

FIGURE 7-3   David Bergman's Flipster lamps have flaps that can be adjusted to change the lighting mood. They are made of 40 percent recycled content. *(Fire & Water Lighting/David Bergman)*

close facsimile. One of Bergman's newest creations is cycLED, which he recently installed over the stylish bathtub of a luxury show house. As the name hints at, the fixture is made from a salvaged bicycle tire hung with recycled glass pebbles. Fitted with white or colored LEDs, the light is playful, and the effect is a curious mix of classic chandelier and funky reuse.

In his bedroom, which is also lit with his designs, Bergman pulled out a couple of dusty lengths of copper pipe fashioned into Steampunk candelabras. "This is how it all started," he explained, blowing off the dust. "I made these candleholders and gave them to friends. This is the original Fire, for light, and Water, for the plumbing pipes. I thought it was interesting because these elements are usually opposites."

Eco-Friendly Materials

A few blocks from Bergman's apartment is the flagship store of the Green Depot, an inviting superstore for environmentally friendly products. In addition to the green light bulb testing station, the shop carries several unique sustainable fixtures. One memorable piece is the Chalkboard Pendant by St. Louis–based artist John Beck (5-inch: $175; 10-inch: $225; 15-inch: $295). Yes, you can write on it with chalk! In fact, customers and staff had left inspirational and humorous messages: "I see green people!" and "Let there be light!" The lamp is made with 95 percent recycled steel.

Other eye-catchers include scaffolding floor ($1,995) and table ($695) lamps by New York–based Rodger Stevens, who makes them from reclaimed exotic woods. Stevens handpicks wood castoffs from demolition sites, piano makers, and elsewhere. The lamps are pretty and whimsical and are designed specifically for CFLs.

There are many different eco-friendly materials that lighting designers can take advantage of. A few main types include:

- Reclaimed or salvaged parts
- Recycled plastics, glass, and metal
- Composites made from waste fiber (paper, clothing, agricultural residues) or bio-based materials (corn or soy, although it's true that these crops can be raised with pesticides and other intensive inputs in some cases)

- Sustainably harvested wood or plant material, such as cork, bark or bamboo (best is third-party-certified, especially by the Forest Stewardship Council {FSC])

When Thomas and Sarah Crowell built their gorgeous home in bucolic Chatham, New York, in 2006, they knew they wanted something sustainable because Tom is director of outreach and resource development for the Columbia Land Conservancy. For their builder, the Crowells chose Connor Homes, a Middlebury, Vermont–based firm that specializes in "the new old home"—traditional early American architecture in dwellings that are prebuilt in a mill and then assembled on site. This reduces waste and minimizes disturbance on lots. It also keeps costs lower. The homes are built to last a long time, according to Connor Homes, which even works with an Amish timber framer.

When it came to their light fixtures, the Crowells went almost exclusively with reclaimed (Figure 7-4). They picked up some from eBay, purchased a few from a local architectural salvage store, and reused some from a Greenwich Village brownstone that had been in Sarah's family. Several of the fixtures had interesting stories behind them, including one from an old New Orleans hotel.

This is one of the advantages of reclaimed materials: the history. One of the authors has an unusual hanging fixture in an entryway that was salvaged, years ago, from a Syrian harem. It works great with a couple of CFLs, and it's a wonderful conversation starter for guests (Figure 7-5).

Some designers claim that copper is a "green material," and it is seen commonly in both indoor and outdoor fixtures. The metal certainly has its appeal, with its beautiful color and, more recently, its association with the trendy Steampunk movement. Copper is durable, long-lasting, and highly recyclable, all of which are eco-friendly characteristics. However, most copper has only limited recycled content, and extraction of new ore is an extremely intensive process using a lot of energy and leaving big scars in the environment. For his part, David Bergman has been backing away from copper in favor of materials that he considers more eco-friendly overall.

For lampshades, search for natural materials, such as linen, decorated prints on recycled cardboard, or felt. There are also great examples of attractive recycled plastics, metal, and glass (Figure 7-6).

Figure 7-4   A view of the Crowell's dining room in Chatham, New York, showing elegant reclaimed fixtures. *(Jim Westphalen Photography)*

Funky Recycled Fixtures

Many artists have made striking fixtures out of repurposed trash, from sea glass to old bicycles (Figure 7-7). One Etsy crafter offers a surprisingly colorful and pretty chandelier made from old Epson ink cartridges ($150, from Lowell & Louise) (Figure 7-8). When it comes to radical reuse, almost anything goes, as long as you are mindful of any potential fire hazards from bulbs that heat up (another good reason to use cooler CFLs or LEDs over incandescents).

FIGURE 7-5   This vintage fixture was salvaged from a harem in Syria. It looks great with CFLs. *(Photo by Brian Clark Howard)*

FIGURE 7-6   This colorful lampshade was made from used snack bags and candy wrappers by artisans in rural Bangladesh. It is available from tenthousandvillages.com. *(tenthousandvillages.com)*

FIGURE 7-7 David Bergman's unique cycLED fixture is made with reclaimed materials. *(Fire & Water Lighting/David Bergman)*

FIGURE 7-8 A chandelier made out of recycled printer cartridges, available at Lowell & Louise. *(David Winton/LowellandLouise.etsy.com)*

Another fun example is the Traffic Light Wall Lamp by Kellen Bain of Elgin, Illinois. As he explained on his blog, Kellen's Metal Art, "When I picked up these lenses I was surprised to learn how blue the 'green' lens was. I decided to replace a boring wall lamp I had in my stairway with the blue lens. . . . I used a compact fluorescent bulb with a cool spectrum to keep the blue look of the lens." Bain also turned a full traffic light into a floor lamp with a switch to control each color.

Nontoxic Finishes

Over the past few decades, awareness has been building about the possible dangers lurking in our homes and workplaces from the volatile organic compounds (VOCs) that off-gas from carpets, vinyl, paints, stains, finishes, upholstery, and hundreds of other products. According to the U.S. Environmental Protection Agency (EPA), indoor air is two to five times more polluted than what's outside as a result of this off-gassing, coupled with low circulation. Little is known about the cumulative effects of living with all these toxins, although scientists think that children, the elderly, pregnant women, and the immune-compromised are most at risk.

It's important to note that lighting fixtures aren't the biggest contributors to indoor VOCs (look to carpets, paints, and large furniture items first). But anything we put in our homes will affect the air we breathe for at least a third of our lives—the approximate time we are there. Therefore, when possible, look for hardware that's treated with finishes that are advertised as having low VOCs and/or that are water-based, which will off-gas less than petroleum solvents. This isn't yet a common selling point when it comes to lighting, but expect it to become more talked about in the coming years.

Remember, too, that using recycled and vintage materials will cut down on off-gassing because the older a material is, the longer it has had to release breakdown products and reach a more stable state. (Think how that "new car smell," which is actually VOCs, drops over time.)

When Kim Lancaster designed her green home in Rhode Island, she used a no-VOC wiring system, which is also designed to be easily upgradable as technology or family needs evolve. The three-inch flexible conduit that supports the wiring is also made with low-VOC material.

When asked about the materials in LEDs, Roy Burton of Dialight said, "The stuff we use for fixtures is pretty harmless. We use reground plastics where we can for housings or even reflectors. That's a good cost saving, as well as being green. We're always looking at ways to recycle and save energy."

Recyclability of Fixtures

The Smart Car has body panels that can be easily removed and recycled and then replaced with fresh ones, which makes owning the car easier as well as greener. Increasingly, lighting designers are also giving more thought to the complete life cycle of their products, including what happens to them when people are done using them. Products can be made to be much easier to break down into recyclable parts. This is becoming standard practice in parts of Europe, particularly Germany, where product "take-back" laws require manufacturers to recycle spent items.

Bergman says that he intentionally designs Fire & Water fixtures to be "noncomplex," meaning that they can be separated and sorted easily for reuse or recycling. Most methods of attachment between materials are mechanical, which makes them easier to take apart than if they were glued (think of most IKEA furniture).

When it comes to LEDs, there has been less of a concern with recyclability because the lights last so long (40,000 hours or more). Eventually, of course, a fixture will outlive its usefulness. "Eventually we will see standardized LED sockets so the LEDs can be replaced or upgraded," suggested Bergman. "From an environmental point of view, I hate the idea of throwing out the fixture with the bulb."

Reduce Materials

Perhaps the easiest way to make your lighting greener is to use less of it—only this time we aren't talking about energy. When Kim Lancaster and Joe Hageman built their green home, their wiring guru, Jeff Mitchell of Rhode Island's Robert Saglio Audio Video Design, used a centrally located utility room in the basement for the core wiring. As a result, he was able to eliminate 80 feet of wire from each of the 62 main runs. Mitchell and his clients also decided to run only as much wire as actually was needed for each device, cutting out the excess that is normally added by most installers. Overall, the project used about 50 percent less wire.

The lesson here is that with smart planning, you can use less wire and fewer fixtures and other materials, which will save you money as well as the earth's resources. Another example is the fact that Philips RoadStar LED streetlights are so efficient that they can be spaced farther apart than conventional streetlights. This saves on materials, as well as capital costs. Further, using low-voltage direct-current (dc) lighting also can lead to use of fewer materials because systems are simpler and everything lasts longer.

Packaging

Much lighting equipment is, naturally, quite fragile, which means that it can require considerable packaging. Some boutique firms are starting to use more recycled packaging materials, as well as bio-based filler, such as cornstarch packing "peanuts." Not everyone is totally pleased with these products, however. David Bergman had to give up on them after they attracted rodents. During a recent interview with *The Daily Green*, executives from UPS said that they have seen problems with bio-based materials, claiming they don't work as well as Styrofoam and melt when they get wet. Still, they can be a viable option in a number of cases.

In general, consumers and businesses can do more to reduce the packaging they use, which should save them money in the process. It's also easy to reuse packaging, and this can help to cut down on waste. If you are buying supplies at retail, bring your own reusable bags to take them home in. If you are ordering from an online retailer, you may be able to request recycled packaging or at least *offset* the shipping impacts by donating a buck or two toward planting trees. Many retailers offer the service via a single click.

Manufacturer Operations

David Bergman has extended sustainable thinking to every aspect of his business. He has reduced paper use in his office—by using only e-faxes, for example—and discourages customers from requesting printed catalogs. He collects any food scraps for composting and buys green power from his local utility.

Lighting companies, just like any business, can benefit from going greener across the board, saving energy, water, and materials. This is not just environmental stewardship—it's good business sense.

Transportation

With rising fuel costs and increasing concern about carbon emissions, people are starting to take a closer look at the impact of transporting so many goods back and forth across the planet. Today, lighting isn't an area that has seen much attention when it comes to sourcing parts locally, but this may change in the future. Right now, even though most LEDs and CFLs are made in China, there is a fair amount of assembly that takes place in the United States and in other areas, particularly when it comes to middle- and high-end fixtures.

One example of a locally produced piece is the hand-blown glass pendant lamp ($995) by Brooklyn artist John Pomp, made exclusively for the Green Depot. The beautiful, dramatic hanging fixture is designed specifically for CFLs to produce a soft, diffuse glow.

Even if local sourcing isn't an option, companies could consider offsetting their operations or using biofuels such as biodiesel in trucks and other vehicles. Every little bit helps!

Make Your Own Lamps

Most lamps are actually pretty simple affairs when it comes to their engineering. All you really need is the proper setting for a bulb, the power supply, and some kind of shade or diffuser to control the light. Hardware stores carry all these basics, and it isn't difficult to flex your own creative muscles. Of course, do be mindful of the heat that certain bulbs can produce because you don't want to increase the risk of fire. (And make sure to take care with all electrical wiring!)

To make it even easier, check out the Westinghouse Make-A-Lamp Kit ($11.95), which comes with a 10-inch brass-plated detachable harp, eight feet of 18-gauge UL-listed cord, a push-through socket, a finial, and all necessary hardware. Combine this with an old milk jug, glass block, or anything else that you can scrounge up, and you've got your own unique fixture (Figure 7-9)! Now that's locally produced and recycled!

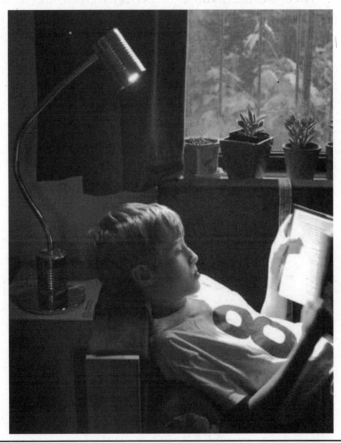

FIGURE 7-9 Learn how to turn used tin cans into a LED reading lamp at Instructables.com. *(Katherine Belsey)*

Summary

We have seen that environmental concerns don't stop at energy use when it comes to lighting, even though that's often where the public debate drops off. There are many other considerations, including the sustainability of the materials in the fixture, transportation costs, and recyclability and reuse.

No one expects you to be perfectly green, and in fact, there is probably no such thing. By giving a thought to the total life cycle of your lighting, though, you can begin to make inroads, most of which also will save you money and hassle in the long term.

CHAPTER 8

Lighting Best Practices and Daylighting

While it's possible to choose a lighting scheme based on detailed calculations of square footage, bulb output in lumens, angles of light, and other factors, it's also true that lighting is highly intuitive. For the most part, it's relatively easy, and affordable, to change the lighting in rooms. Simple adjustments include swapping out bulbs and shades, adding or subtracting floor lamps, or changing your curtains. In other cases, more intensive changes may be needed, such as rewiring fixtures or adding a window or skylight to let the free sunshine in.

In general, it's best to light rooms with a number of sources rather than from a single lamp, whether it is an overhead or based on the floor or wall. By using several sources, you help to reduce glare and harsh shadows and have more flexibility.

Let's take a look at some ways to maximize comfort and productivity, as well as energy savings, in your home or business.

Indoor Lighting Strategies

Here are some general tips for smarter, more efficient lighting:

- Use focused task lights as much as possible, reducing the need for general ambient lights.
- Use the lowest-wattage bulb you can for a given task.

- Turn off lights when you aren't using them; sensors, timers, and other controls can help.
- Use light colors on walls, which will reflect more light and reduce your need to generate it.
- Maximize the use of windows and skylights, which is called *daylighting*.
- In retrofits and new construction, make sure that all lighting meets, or exceeds, applicable codes. This will improve safety, preserve your property value, maintain efficiency, and reduce your liability and risk of fines.

Know When to Turn Off Your Lights

Naturally, the surest way to save energy with lighting is to reduce the length of time that it is switched on. However, the actual cost-effectiveness of when to turn off lights depends on a number of factors, chiefly the type of light and the price of electricity. Turning off incandescent lights starts saving money after just a few seconds because they don't require much to start up. Also, since incandescents are not appreciably degraded by turning them on and off, they can be switched on and off frequently without loss of life.

With fluorescents, on the other hand, it's a bit more complicated. For recent fluorescent technologies, it takes only about five seconds of use to use the same amount of power that it requires to start up, according to the U.S. Department of Energy (DOE). However, since frequent switching does decrease the life of fluorescents, experts recommend shutting them off only if you won't be using them for 15 minutes or longer.

To calculate precisely how much energy you are saving by turning a light off, follow these instructions from the DOE: First, determine how much energy the light consumes when on. Do this by locating the watt rating printed on the bulb. Multiply that number by the number of hours the light is on, and divide that by 1,000 to convert to kilowatt-hours. For example, for a bulb rated at 40 watts and used for 1 hour, this will consume 0.04 kilowatt-hour—or it will save 0.04 kilowatt-hour for every hour it is off.

Next, find out what you are paying for electricity. For some customers, this may vary based on time of year and peak or off-peak periods, so you can do the calculation multiple times to get more exact results. Look at your electric bill to see what the utility charges per kilowatt-hour, and multiply the rate by the amount of electricity saved by not using the bulb. With this example, if your electric rate is 10 cents per kilowatt-hour, the value of the energy savings would be 0.4 cent ($0.004) per hour. Of course, the value of the savings will increase with higher-wattage bulbs and as electric rates go up. If you know the cost of a replacement bulb and any labor costs changing it would require (for commercial managers), you can estimate when it is most cost-effective to make sure that lights are shut off.

Avoid Overlighting

As in most of life, too much of a good thing can be a problem. Too much lighting not only wastes energy, but it also can be uncomfortable to room occupants. No one, except maybe the occasional rock star or retro new waver, wants to wear sunglasses indoors. Overlighting can cause excessive glare and actually can make it harder to complete tasks.

Outdoors, overlighting can become light pollution, which disrupts wildlife, disturbs neighbors, and impedes the view of the starry sky.

Choose the Right Light for the Job

Low-pressure sodium lights may be efficient, but their poor color rendering and long startup times make them unappealing for most indoor spaces. Similarly, many cheaper compact fluorescent lamps (CFLs) have low color rendering indexes (CRIs) and harsh color temperatures and may not be ideal for living spaces, although they can work fine in utility areas, hallways, and the like. If you replace a 60-watt incandescent bulb with a 14-watt CFL and you don't like the color or look of the light, try a different CFL. Sometimes, using a higher-wattage CFL will help to compensate for the difference, since it will produce more light.

Artists and crafters have known for a long time that full-spectrum and high-CRI lights can help them to better see colors and details. This is also why paint mixers often work under full-spectrum lights—to better see the "true color" of the pigments. Grow lights for indoor plantings are also often full spectrum or nearly so to maximize the benefits to vegetation. Similarly, anyone who has ever kept reptiles knows that the animals are healthier if their caretakers use lamps that are formulated to more accurately replicate natural environments.

Along these lines, some manufacturers tout full-spectrum lights for phototheraphy, which is used primarily to counter seasonal affective disorder (SAD), a type of depression that seems to result from reduced exposure to sunlight during the winter. However, the role of artificial lighting in reducing SAD is not well understood, and some research suggests that white or blue light (or just simply the amount and timing of light) may be the crucial part of prevention.

Lighting Room by Room

The lighting needs of each room can be a bit different depending on how the space is used primarily, as well as its size, orientation, color, and relationship to other rooms. Still, there are some general guidelines that can help to get you started:

- *Entryways.* Ideally, lighting in foyers should offer a transitional level of brightness between the outdoors and indoors to help welcome visitors and guide them further inside. Notice that many fine apartment buildings and hotels use indirect lighting in entranceways, creating a soft, warm air of elegance. Wall sconces and overhead lights work well together. In private homes, these lights are rarely left on for long periods, so in that case they often aren't good candidates for fluorescents.
- *Kitchens.* The kitchen is the heart of a home, so it should get lots of light. It also is a prime candidate for efficient technology because people spend a lot of time in their kitchens, and those lights are among the most used. A central fluorescent overhead fixture often works well for ambient lighting, paired with some additional features. Keep in mind that higher ceilings will demand brighter light bulbs, as will darker colors, including dark marble countertops.

  Undercabinet lighting is becoming increasingly popular and is a great way to focus light right where you need it—on the counters where food is prepared. Light-emitting diode (LED) strip lights work well for this, as do small clusters of bright white LEDs made especially for this purpose. Fluorescent undercabinet fixtures are also widely available, and the housings tuck away neatly out of sight. Undercabinet lighting can make a kitchen feel bigger, as well as more contemporary. Another popular place for indirect lighting is at the tops of cabinets, which can give the room a soft glow, as well as highlight artwork. Don't forget to make sure that any fixtures that could come into contact with water are protected by a ground-fault circuit interrupter to prevent electric shocks.

Another way to add interest to a kitchen is with hanging pendants, many of which are quite colorful and which often work great with CFLs. Pendants also can be placed over high-use areas to provide more light when needed, or they also can be used to divide up, as well as decorate, different sections. They are often used over eating areas.

Ceiling fans are popular in kitchens, as well as in bedrooms and other areas. Note that fan vibrations can decrease the life of CFLs, so it can be a wise investment to get models that are designed for the purpose. You also can choose Energy Star–registered fan-light combos, many of which are specifically designed for CFLs.

- *Dining rooms.* Traditionally, the centerpiece of a dining room is a large hanging light, whether it's a classic crystal chandelier or a bold avant-garde statement piece. For the best effect, the fixture generally should be at least 12 inches narrower than the table and should hang a minimum of 30 inches above the table's surface to avoid glare, not to mention collisions with taller guests. As with all rooms, light should not be from only one source; a chandelier should be augmented by one or more adjustable downlights or wall lights.

- *Living rooms.* Think layers of light, which you can apply in succession as needed. This can include some overhead ambient lighting, task lighting for reading or other activities, and accent lights—such as recessed low-voltage fixtures—to add a sense of depth and show off parts of the decor (Figure 8-1).

- *Bathrooms.* Bathrooms also should be well lit, although, unfortunately, many aren't. It's usually easiest to install a single light above the mirror, and this is what builders often do in apartments or older homes. But this scheme produces harsh shadows and high glare. It's much better for putting on makeup or shaving to

FIGURE 8-1   Note the valance lighting (via dimmable fluorescents) in this living room by David Bergman. The pendants in the background are antique glass fitted with CFLs. *(Fire & Water Lighting/David Bergman)*

have lighting coming from either side of the mirror, ideally 35 to 40 inches apart and at head level (think about the dressing rooms of stars shown in classic movies). Pair this vanity light with an overhead, ideally something built into an exhaust fan unit.

- *Bedrooms*. Soft, warm light is best for bedrooms, and dimmers are particularly effective. With overhead lighting, reduce harshness and glare by using recessed downlights that focus toward the foot of the bed or toward the sides of the room. Pendants and wall scones also can add beauty and comfort. Pair these with table or directional lighting for reading on either side of the bed, as well as task lighting for closets.

People want their bedrooms to be cozy, and many are reluctant to consider CFLs because of their past associations with offices and cool temperatures. This is where new extrasoft CFLs shine and where it pays to use a good fixture that moderates the light. "People generally don't like to look at CFLs. I mean let's face it, they aren't the prettiest looking things," Bergman explained as he showed off the diffusing fixtures in his own bedroom. "So I make sure you can't see the bulb." Bergman's bedroom is decidedly cozy and is lit with CFLs set in his fixtures.

- *Offices.* Since most offices—including home offices—are frequented for many hours at a time, fluorescents pay for themselves rapidly. Fluorescents also provide bright light that facilitates good visual acuity. To decrease reflections, avoid placing ceiling fixtures in front of a desk. Instead, light should come over the shoulders of the desk's occupant. For more natural, comfortable lighting, use the fluorescents in indirect settings. Pair them with wall and accent lighting for variation, and provide task lighting as needed.

- *Corridors.* Hallways should be neither a lot brighter nor dimmer than adjoining rooms so that moving through them is not jarring to the eyes. Yet they should have enough illumination for safety, particularly in commercial settings. One general rule of thumb is to place a fixture every eight to 10 feet, either wall- or ceiling-mounted. Fluorescents aren't usually ideal in private residential hallways because the lights don't need to be on much, although they can make a lot of sense in commercial buildings. LED strip lights can work well in nearly any hallway, as the Hageman family discovered in their green Rhode Island dream house.

- *Stairs.* For obvious safety reasons, stairs must be well lit, particularly in commercial buildings, where liability is a serious concern. David Bergman established a welcoming, diffuse light in the exposed-brick stairwell of his Manhattan condo building by placing a large, wall-mounted CFL fixture from his Parallel Universe series at each landing. Large commercial buildings are also increasingly installing recessed LEDs low to the ground, which illuminate each step. In typical homes, it's common to use a single close-to-ceiling fixture or a chain-hung fixture at the center of the stairway.

- *Basements.* Good lighting can produce surprising transformations in basements, which often seem more dank, dreary, and inaccessible than they need to be. If possible, drive sunlight into your basement. If you can't put in a few windows, consider tubular skylights. For artificial lighting, fluorescent ceiling fixtures provide a lot of light for low cost. To make your basement seem larger, light an entire wall with recessed lights mounted on the ceiling. The lights should be spaced at an equal distance from each other and the wall. If you have a workbench, desk, or sewing machine in your basement, use task lighting, as you would in other areas.

- *Accents.* If you have a fine painting, family portrait, or other type of wall covering you want to highlight, consider a dedicated picture light or recessed low-voltage light. This can be especially impressive in formal dining, living, or sitting rooms or in corporate lobbies. Recessed low-voltage lights or LEDs can beautifully illuminate china cabinets and hutches. If you want to emphasize or set something off, such as a table, buffet, or artwork, place wall fixtures on either side of that element. Place a light on a wall behind a plant to set off its silhouette.

Outdoor Lighting

We may not necessarily give it as much thought, but outdoor lighting is an important part of property management, and it provides safety, security, and aesthetics. Studies are inconclusive on whether outdoor lighting actually decreases crime, although most security experts believe that motion-tripped lights probably help. Unfortunately, outdoor lighting also results in significant energy use, and it can contribute to light pollution, which can disrupt wildlife and the sleep cycles of other people.

As stated earlier, it's important not to overlight your yard (you also can look for fixtures that have been certified to reduce light pollution by the International Dark Sky Association). Outside, a little light can go a long way. Focus on the areas where you actually need illumination (driveways, pathways, and porches), and then add one or two areas to highlight, such as a sign or attractive tree or key part of the building. Remember to use controls to maximize utility but minimize energy use (Figure 8-2).

Here are common outdoor lighting strategies:

- *Uplighting.* Lights are placed at ground level and are aimed up toward a focal point, such as a sign, wall, or tree. The fixtures are often floodlights or spotlights.
- *Downlighting.* As the name implies, elements are lit from above, often with spotlights or floodlights.

FIGURE 8-2   Install attractive and functional outdoor lighting and increase security and safety, as well as show off landscaping and architectural elements. *(Nightscaping)*

- *Spotlighting*. A strong beam focuses on an object, such as a flagpole.
- *Pathlighting*. Lights are placed low to the ground to illuminate a path or a driveway.
- *Backlighting*. Lights are placed behind objects, such as plants, and fixtures are concealed.

Here are some additional outdoor lighting tips: Installing outdoor lighting can require some of the same thinking that goes into other aspects of landscape design, especially the need to keep future changes in mind. "Install conduit under driveways or patios before paving or bricking, and have the ground-fault circuit interrupter (GFCI) receptacles installed before getting started," Randall Whitehead, a landscape lighting designer in San Francisco, told *Popular Mechanics* magazine. He added, "Buy fixtures with more wattage capacity than you need, [and] then increase wattage in the future by replacing smaller wattage lamps with higher-wattage ones [within the capacity of the transformer] as the plants mature." To avoid a monotonous look, mix and match fixtures and spacing.

Lighting Maintenance

Over time, light levels can fall because of

- Fixture dirt
- Lamp aging

These factors can reduce light levels by as much as 50 percent, even though the lights still may be drawing full power. This is why it is a good idea to keep fixtures, lamps, and lenses clean. Most experts suggest that lighting should be cleaned every six to 24 months.

It also helps to thoroughly clean or repaint small rooms every year and larger rooms every two or three years. Again, lighter walls require less light.

Daylighting

As we learned in Chapter 1, studies have shown that introducing daylighting into buildings can boost comfort, morale, and productivity, as well as decrease absenteeism. So what is daylighting? It's really a technical-sounding term for something that couldn't be more intuitive. Simply put, *daylighting* is the use of windows, skylights, and advanced features to bring natural sunlight into a building. Not only is sunlight free, but it is also full spectrum, so it offers perfect color rendering and other positive features (Figure 8-3).

If you are worried that daylighting will conflict with heating and cooling goals, note that there are ways to minimize this. For example, today's energy-efficient windows can be twice as efficient as windows sold just 10 years ago. Overall, savings from daylighting can cut lighting energy use by 75 to 80 percent, according to the DOE. The DOE has found that many commercial buildings can reduce total energy costs by up to one-third through daylighting.

Here are some points to keep in mind:

- As with solar panels, south-facing (equatorial) windows typically are best for daylighting because they allow the most winter sunlight into the home but little direct sun during the summer, when heating is a concern.

FIGURE 8-3 Extensive daylighting makes the lobby, cafeteria, and galleries of Hearst Tower in New York City memorable, as well as less expensive to light. *(Photo by Brian Clark Howard)*

- North-facing (pole) windows also work well for daylighting because they admit relatively even light, with little glare, and they also don't add much to summer heat gain.
- East- and west-facing windows provide strong daylight in the morning and evening, although the low angle of the rising or set-

FIGURE 8-4   Daylighting through the roof and walls brightens up this wood products factory in Costa Rica. *(Photo by Brian Clark Howard)*

ting sun can be distracting. These orientations also can cause glare, and they can let in a lot of heat in mornings or evenings during the summer (Figure 8-4).

All windows will let in light, of course, although high clerestory windows work especially well (see Figure 8-5). To maximize the light from any window, have it face a light-colored wall to reflect and diffuse the light into your space. You also can add a large white sill to the window, which will help to reflect the rays.

To reduce heat exchange through windows, use blinds and curtains to block the strongest rays when it's hot or hold more heat in when it's cold and the sun is down (Figure 8-6). Awnings and overhangs also can help, as can recessing the window. Property owners can plant trees or bushes in front of windows to keep them cooler or employ temporary external screens. Of course, the better your windows are, the less of an issue heat loss or gain will be. Double- and triple-glazed options have a strong insulating effect.

If you want to save more, consider upgrading your windows with a low emissivity (low-e) film or coat. This relatively simple technology can reduce heat loss or gain through old windows by up

FIGURE 8-5   Massive clerestory windows light up this award-winning "green" Walmart superstore in Aurora, Colorado. Studies have shown that such daylighting improves employee productivity and attendance and raises customer sales. *(Photo by Brian Clark Howard)*

FIGURE 8-6 Use adjustable blinds, instead of artificial lighting, to control light and shadows through the day. *(Photo by Brian Clark Howard)*

to 40 percent. The microscopic metal oxides in the film can block solar energy (and harmful ultraviolet rays) from entering your home, whereas the opposite side helps to reflect heat back in.

FIGURE 8-7 This multistory apartment building in San Jose, Costa Rica arranges interior rooms around small courtyards in order to bring in natural light. *(Photo by Brian Clark Howard)*

Skylights

Another way to let the free sunshine in is with skylights, which have come a long way over the past few decades (Figure 8-8). New designs seal much tighter against leaks and heat loss. If you are looking to put a new one in, choose an Energy Star–rated system.

Skylights are perhaps most common in great rooms and porches, although they can work well almost everywhere, from bathrooms to kitchens to bedrooms. They also work wonders in commercial spaces, where they not only result in substantial energy savings but also create a pleasant ambiance. Tubular skylights work particularly well for small rooms such as bathrooms and hallways. They're easier to install, less expensive, and more energy efficient than traditional skylights. Some installers suggest that the optimal size of skylights is four to eight percent of floor area.

For buildings that can't have traditional skylights or for people who want to stretch the effect further, there are *solar tubes*. These

FIGURE 8-8   Skylights help cut down on use of artificial lighting, and they tend to make for more attractive indoor spaces. *(Photo by Brian Clark Howard)*

devices mount on the roof and use advanced optics and special duct-work to direct sunlight deep into the building, bouncing around cor-ners and spreading out over large internal spaces. Solar tubes can be simple, with a reflective interior coating, or they can incorporate light-conducting fiber optics. They are often capped with a trans-parent dome "light collector" and terminated with a diffuser to spread out the light in the building. Since they have a smaller area exposed to the outside, they have less of a problem with heat gain or loss. California-based Solatube International is a leader in the field, offering solutions for home and business clients (Figures 8-9, 8-10, and 8-11).

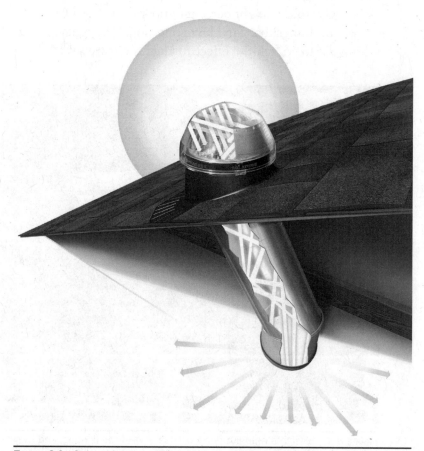

Figure 8-9 Solar tubes can help to bring sunshine deeper into buildings. *(Solatube International)*

FIGURE 8-10   A bathroom before and after a Solatube was installed. *(Solatube International)*

FIGURE 8-11   The diffuser of a Solatube, which spreads daylight deep into buildings. *(Solatube International)*

Advanced Daylight Harvesting

Daylighting also can be incorporated in lighting controls in a process called *daylight harvesting*. The basic idea is that photosensors can measure the amount of light coming in and adjust artificial lighting accordingly.

"If you're at a workstation near a window, do you need your lights full on? Not on a sunny day," said Michael Smith of Lutron. The company's lineup includes electronically operated shades that can "talk" to the control system wirelessly. "Our system reads the sun and clouds, and maybe the shades respond, or lights are dimmed," explained Smith. "The energy savings can be substantial."

Summary

Lighting is largely intuitive, and it is relatively easy to experiment with different scenarios. Still, there are some general rules of thumb that can help to guide you to effectively and efficiently light each space. Half the battle is choosing the right tool for the job, and keeping everything in good working order. When everything works together, truly wonderful things can result. Figure 8-12 shows a home in which a renovation was completed in a sustainable and energy-efficient manner.

Figure 8-12 This 1960s California ranch in Silicon Valley was recently renovated by David Bergman. Sustainable features include bamboo cabinetry with non-formaldehyde-containing interiors, cork and linoleum floors, recycled glass counters and tiles, natural clay finishes, dual-flush toilets, and energy-efficient lighting (primarily CFLs with dimming and daylighting). *(Fire & Water Lighting/David Bergman)*

CHAPTER 9

Solar and Next-Generation Lighting

We have seen that green lighting technologies are evolving rapidly, thanks to considerable interest from the public and investment from governments. A number of promising new technologies are also being investigated that could provide even greater flexibility and benefits in the near future.

Another exciting area is solar lighting, which at first may sound like an oxymoron but actually can work very well. With a little device called a *battery*, solar lighting can provide illumination without being tied to the grid by taking advantage of the energy of the sun. Let's take a closer look.

Solar Lighting

On the Internet, as well as in gift, educational, and green stores, you can find an increasing array of solar-powered products, from backpacks that charge cell phones to a solar-powered pith helmet with a built-in cooling fan. There are also a number of solar-powered lighting gadgets, such as table lanterns, flashlights, and reading lights (Figure 9-1). Some of these products work pretty well, especially light-emitting diode (LED)–based solar lights, although it's true that some of these gadgets are of questionable value (pith helmet anyone?).

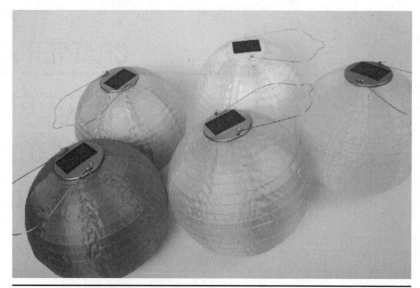

FIGURE 9-1   These Soji solar-powered lanterns make great accent pieces.
(Bambeco)

One potential issue, according to inventor and materials science PhD Saul Griffith, is that slapping a solar panel on a product that won't be used much isn't really a good use of resources. It takes a considerable investment of materials and energy to make a solar panel, and it normally takes a few years for a panel to "pay back" the amount of energy that was required to produce it—and this is with constant operation. "Solar panels used on many consumer devices are never going to pay for themselves in terms of energy," Griffith said at the 2009 Greener Gadgets conference in New York City. Even so, a number of the other panelists at the conference, including Greener Gadgets cofounder and *Inhabitat* publisher Jill Fehrenbacher, argued that the educational value of getting solar panels out in front of the public, as well as support of the industry, made up for the energy deficit. Still, it's important to remember that having a solar panel doesn't automatically make something green: We really should be looking at the total life cycles of things, as well as the usefulness and durability of the product, among other factors.

One small solar product that can make a big difference is the BoGoLight, so named because for every flashlight bought by someone in the developed world, one light is sent to people in need in the developing world. The handy LED flashlights are powered by embedded solar panels, and they last for years—for thousands of hours

FIGURE 9-2   For every BoGoLight you buy, SunNight Solar donates one to someone in need. (SunNight Solar)

of illumination (Figure 9-2). The nickel–metal hydride batteries need only be replaced about once every two years. Mark Bent, founder of BoGoLight maker SunNight Solar, is a former U.S. Marine and diplomat who has lived all over the world. He was struck by the fact that one-third of humanity still uses kerosene for lighting, even though the fuel is dirty, polluting, and aggravates asthma, especially in children. It is also relatively expensive. "Many people in Africa and India spend 30 percent of their income on kerosene," Bent said at the Greener Gadgets conference in 2009.

"A top Ugandan diplomat in Washington, DC, got there because his parents had saved money for him for kerosene, so he could read," added Bent. "He won a scholarship for grade school, then for high school and college. His friends that he grew up with in his village are herding goats now. Hopefully, my light can bring about change like this." But Bent also acknowledged that he is often surprised at the ways in which his lights help people. "I asked a gentleman in Eritrea what was the best value for my light," Bent added. "I thought he would say security, health, or reading. But he said he had more baby goats. It turned out he could help them with troubled births because he could see. He didn't have to go get a kerosene lamp, which took too much time and caused him to lose more goats. Those people measure their wealth by the number of goats they herd." Houston-based SunNight Solar (www.bogolight.com) also provides BoGoLights to victims of disasters, such as the catastrophic Haiti earthquake.

Bent said that he has researched making BoGoLights out of recycled or bio-based plastics, but his engineers have not found a solution that they think is rugged enough for daily use in remote areas. "We're a 'three-p' company: people, planet, profits," said Bent. "Soon people will start asking: 'Why am I buying things that don't make any sense, that don't have sustainability built in right from the front?'"

Solar Outdoor Lights

Another grid-free product is the solar-powered umbrella, which is offered by a number of companies. This functional patio accessory has a small built-in solar cell that charges batteries during the day. When night falls, a photosensor switches on LEDs embedded in the underside of the umbrella. It's a convenient, attractive way to light a table for the evening or serve as accent lighting.

The same idea has been applied to LED holiday lights. Online solar superstore Real Goods offers strands of LED lights and wreaths (made with recycled plastic faux evergreen boughs!) with LED lights that are powered with their own small solar panel. These products also have built-in photosensors.

In fact, solar outdoor lights are becoming increasingly common because prices keep dropping and because they are much easier— and safer—to install than traditional lighting. With solar-powered

FIGURE 9-3 A solar-powered stop sign in the parking lot of a "green" Walmart in Colorado. *(Photo by Brian Clark Howard)*

lights, there are no wires to fool with or that need to be buried or strung between landscaping. All you have to do is position the light where you want it, point the solar cell toward the sun, and make sure that your sensors are set correctly. You should never have to think about your lighting again—at least not for a few years, when a battery may need replacing or a panel may need a good cleaning.

It's true that not everyone is beaming about solar patio and accent lights. We recently met with the chief product engineer for a major lighting company, and he was critical of the technology (which competes with his business). "Solar lights work okay at first, but they start losing light quality pretty quickly, whether the solar panel gets covered or damaged, and most of them are, let's face it, pretty cheap-looking affairs that I wouldn't want in my yard," he said.

True, there are some discount solar lights on the market that wouldn't look right in front of the Ritz Carlton and that may not live up to some marketing claims. But that's true of any type of product. There are also high-end solar solutions that have earned rave reviews. For example, check out the beautiful solar LED accent fixtures in pewter and other fine metallic finishes (Figure 9-4) that are available from GREENCulture (www.eco-lights.com). In all cases, it's a good idea to try before you buy, and make sure that you follow the manufacturer's directions.

Outdoor solar lighting will work in most parts of the world as long as you aren't too near the poles. Remember, though, that any given solar product will only work well as long as the solar cells receive the manufacturer's recommended hours of sunlight. The ad-

FIGURE 9-4   Solar accent lighting at twilight. *(Photo by Brian Clark Howard)*

vertised nightly run times for most products are based on specific sunlight conditions. If they get less sun, the lights will be on for less time. And yes, cloudy days can decrease the amount of energy that can be collected, as can shorter days during the winter. But these issues can be managed by properly sized and sited fixtures.

Do watch out for shading of the solar cells by vegetation or other features, and try to keep them clear of bird droppings. Also be aware that insufficient battery charging can reduce the life of the battery.

Solar lighting systems can be self-contained units, or the solar panel can be separate from the light fixtures, which then can be placed in the shade. They can range in size from tiny accent lights to high-beam security lights. The good news is that solar lighting systems are becoming increasingly available in hardware, lighting, and discount stores, as well as through environmentally oriented mail-order companies.

Types of solar outdoor lights include:

- *Solar stepping-stones.* These popular products work great as accent lights. They typically combine a small solar cell, battery, photosensor, and LED light in a molded resin shell that resembles a natural stone. Just set them out where you want them, and you can move them at any time.
- *Solar garden lights.* The same idea as above, just in different housings. These can be designed as lanterns, posts, or patio fixtures.
- *Solar spot, flood, and security lights.* Spotlights and floodlights can be good candidates for solar power because they are often needed far away from existing wiring. These lights can be paired with motion sensors to make security lights and can be lit by halogens or LEDs. A popular example is the MSL180W from Cooper Lighting ($65), which has a bright 13-watt halogen bulb. OutdoorSolarStore.com sells several designs of solar-powered post lights (starting at $185), as well as "retrofit" kits starting at $125. With these products, you can take advantage of an existing pole, and swap out a conventional lamp with green technology.
- *Solar holiday lights.* Since LED holiday lights use so little energy, solar-powered strands may not make financial or environmental sense in places where plugs are available. But they can work in remote locations or offer flexibility in areas where you don't want to run a wire.

Solar Lights for Developing Countries

It may surprise you to hear that a quarter of humanity is still without electricity. And at current rates of rural electrification, it will take a long time to extend the grid to every last hamlet and home. This is why some experts point to solar lighting as a more affordable and viable alternative to traditional infrastructure in some areas, especially in politically unstable regions. If you currently have nowhere to plug in conventional lighting, then solar lighting can be a bright solution.

Shayne McQuade, the New York–based inventor of the iconic Voltaic solar backpack, is currently working on getting low-cost solar lighting to people in the developing world "funded by First World sales," as he explained in a recent interview. McQuade is currently testing prototypes. "If a solar light costs $100, then that eliminates the three billion people who need it most," McQuade said. His strategy is to make the lights as efficient as possible so that they require smaller (and therefore cheaper) solar panels.

McQuade says that he is inspired by the work of D.light Design, a company that was founded in Silicon Valley in California but which recently moved to New Delhi, India, to be closer to the people it serves. D.light has brought illumination to more than 1 million people around the world, largely in India and Africa, through affordable, bright, efficient solar lanterns. People in the developed world can support D.light's efforts by buying a Nova or Kirin solar lantern, both available from Amazon for $45 or $15, respectively. The lights help developing world families improve their productivity an average of 30 percent, according to D.light, by allowing tasks and studying to be done when the sun goes down.

In the chaotic days following the disastrous earthquake in Haiti, relief and recovery efforts were made easier by solar LED light systems. These have been deployed in Haiti and the Dominican Republic for years, thanks to Calgary, Alberta–based physician Jan Tollefson and her group, Add Your Light Charitable Foundation (www.addyourlight.org). Tollefson works with Jose Rivas, president of Dominican Fundacion Agrega Tu Luz, and others, to provide light to those who don't have it or who don't have reliable service and to help build up small businesses in selling, installing, and servicing solar lighting systems (Figure 9-5).

Recipients of the foundation's lights pay something for them on a sliding scale according to their ability. Tollefson believes that

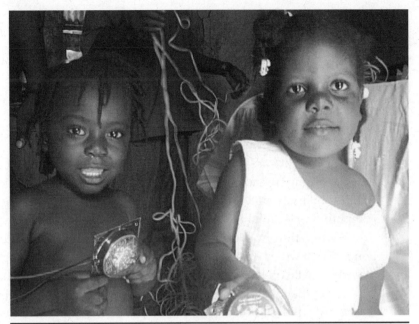

FIGURE 9-5 These girls in Haiti now have lights in their home, thanks to Add Your Light Charitable Foundation and solar-powered LEDs. *(Katherine Buckel)*

this encourages people to take much better care of the units and to actually use them as opposed to if they had gotten them completely free. Each Add Your Light system uses a five-watt solar panel, three custom-built LED lamps, and a 12-volt, seven-ampere-hour rechargeable battery.

Add Your Light isn't the only group in this field. Another benefactor is Palm City, Florida–based Sol, Inc., which donated $400,000 worth of its solar lighting to Haitian relief efforts. The company even shipped systems on a catamaran lent by a wealthy donor. Globally, Sol has installed more than 40,000 solar lighting systems in more than 60 countries on six continents.

Next-Gen Lighting Technology

Lighting technologies are evolving rapidly, thanks to increasing interest in saving energy, reducing emissions, and improving quality. Considerable investments are being made in research, and "breakthroughs" are announced frequently from laboratories around the world. Of course, it's difficult to predict the future and know what actually might catch on, but let's take a brief look at some of the emerging technologies that are currently in development.

Organic Light-Emitting Diodes

There are many efforts underway to improve LEDs using advanced materials and experimenting with new design elements. One particularly "bright" area of research is on silicon-based lighting. Another emerging technology is so-called organic light-emitting diodes (OLEDs). Simply put, if the emitting layer of an LED is an organic compound, it is an OLED.

To work, the organic emitting material must have conjugated pi bonds (don't worry if you don't know what this means). The organic material can be present in a crystalline phase or as a polymer. In the latter case, the result is sometimes called a *polymer LED* (PLED) or *flexible LED* (FLED) because they are often flexible (Figure 9-6).

Supporters believe OLEDs show considerable promise as inexpensive displays and eventually even as integrated luminous panels, wallpaper, or cloth. They are already showing up in the displays of some mobile products. Currently, however, OLEDs have low life expectancies and are less efficient than standard LEDs. However,

Figure 9-6   The Philips luminaire chandelier concept features five OLED "blades" that get brighter when movement is detected nearby. *(Philips)*

they are improving rapidly, and breakthroughs are frequently announced by tech companies. Sony has already come out with an OLED TV, which is expensive now but may be a harbinger of things to come.

One potential issue with OLEDs is that they typically contain the metal alloy indium–tin oxide, and indium is rare, expensive, and difficult to recycle. However, new research has yielded some promising results by replacing indium with a carbon-based material called *graphene*. This flexible material theoretically would allow printing of OLEDs in a process similar to the way documents are printed today.

Quantum Dot LEDs

Another area that has engendered considerable excitement resulted from an accidental discovery, as many new technologies do. A few years ago, a graduate student at Vanderbilt University, Michael Bowers, was working with quantum dots, which are crystals that are only a few nanometers in size (less than 1/1000 the width of a human hair). Normally, when Bowers shone a light on the dots, they emitted blue light. But one day he hit them with a laser, and they produced a radiant soft white glow. So Bowers and another student coated a blue LED with the dots, held in polyurethane, and the result was a pleasant light that was similar to that of an incandescent bulb.

The emission of quantum dots can be tuned throughout the visible and infrared spectrum, meaning that LEDs can be made in almost any color. The technology is being researched further, and it may soon help to improve the quality of LEDs.

Vu1

Another next-gen efficient lighting solution is being touted by Seattle-based Vu1 Corporation, which claims to be on the verge of rolling out its products (the company did not respond to requests for an interview for this book). According to Vu1's Web site, the product will be a screw-in light bulb that works in standard fixtures and is based on "electron-stimulated luminescence," which is described as "using accelerated electrons to stimulate phosphor to create light." As the company points out, this is a process that is similar to what happens inside current fluorescents, as well as the cathode-ray tube displays of older TVs and computer monitors. "Vu1 merged several existing and proven technologies [and] then uniquely adapted them for lighting," claims the company.

Vu1's technology contains no mercury and is estimated to be around two-thirds more energy efficient than current incandesent bulbs (approaching the efficiency of compact fluorescent lamps). It is also estimated to last 6,000 hours, about six times longer than current incandescent bulbs (again, approaching fluorescents). The Vu1 bulb is being marketed as coming in a shape that is more familiar to consumers than compact fluorescent lamps (CFLs), and it is said to have a better color rendering index (CRI), around 95 (closer to incandescents and the sun). The technology also reportedly doesn't require special heat dissipation, unlike LEDs, and it should produce an even glow in many directions, also unlike LEDs. The first Vu1 product to be made available is predicted to be an R-30 reflector bulb, to be followed by other types.

Plasma Lighting (and Lava Lamps!)

Plasma lights were first invented at the turn of the last century by the brilliant Nikola Tesla, who discovered them while running his high-voltage experiments. Plasma lights have been revisited a number of times over the years since and have attracted a fair amount of interest recently. Most plasma lamps are electrodeless, and they work by exciting a plasma in a closed transparent bulb using radiofrequencies.

What's a plasma? Most scientists now consider it a distinct state of matter, alongside gases, liquids, and solids. A plasma most resembles a gas, except that many of its particles are ionized and thus conductive of electricity. This makes plasmas exhibit complex behavior, such as forming filaments. (Plasmas are most common in outer space, where they are thought to comprise the vast majority of matter in the universe.)

You can get an idea of what plasmas look like from the "plasma globes" that were popular novelty and prop items in the 1980s—and that still can be found in museums, science classrooms, and online. Invented by MIT undergraduate Bill Parker in 1970, the globes generate tendrils of purplish light from a central sphere (which is actually an electrode). These types of lights aren't efficient for illumination, and they have some drawbacks, including causing interference with other radio-sensitive devices (including pacemakers).

By the way, in case you are wondering, the popular lava lamps that adorned so many college dorm rooms in the 1960s, 1970s, and beyond aren't really an advanced form of lighting. They actually have an incandescent bulb in the base. The blobs and lenses of mesmerizing liquid are caused by a wax mixture that, when heated, expands and contracts at a different rate than the water in which it is suspended.

But "real" plasma lights are getting another look by lighting professionals, especially to replace metal halides for commercial bright-white applications. Plasma International, for example, markets "sulfur plasma lighting systems," which the group claims emit 75 percent of their output as visible light and only one percent as harmful ultraviolet (UV) light. The lights are said to have a correlated color temperature around 6,000 K and a CRI of 86. They are said to be quite dimmable and don't contain hazardous materials, being made of quartz, sulfur, and argon.

Another purveyor of plasma lighting is Silicon Valley–based Luxim Corporation, which is also targeting the commercial sector. The company claims that its LiFi (for "light-emitting plasma") lighting is full spectrum, dimmable, and efficient for high-illumination applications, such as streets, parking lots, warehouses, and factories. Luxim says that its 266-watt product effectively replaces a 400-watt metal halide or high-pressure sodium unit for a substantial energy savings. LiFi products are rated for 50,000 hours. They produce more light with better color rendering than LEDs (Figure 9-7).

According to Luxim, the LiFi uses a solid-state device to generate radiofrequency energy, which then powers a plasma light source. This is said to combine the high reliability of solid state with the high brightness of gas discharge.

Luxim technology lit up the glamorous night at the 2010 Academy Awards, giving spectators and more than a few paparazzi a clear view of the world's most famous faces. The mobile plasma lights were powered by a clean hydrogen fuel cell (thanks to a partnership with Sandia Labs, Boeing, the California Department of Transportation, and others), and they required only 2.3 kilowatts—less than the typical 4.4 kilowatts.

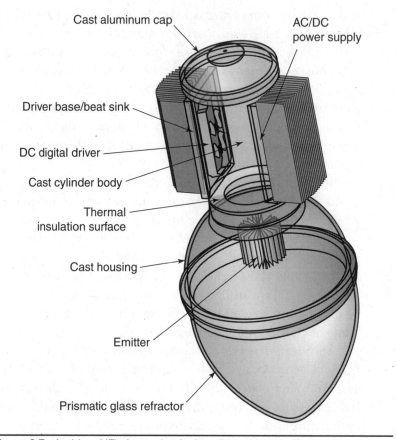

Cast aluminum cap

AC/DC power supply

Driver base/beat sink

DC digital driver

Cast cylinder body

Thermal insulation surface

Cast housing

Emitter

Prismatic glass refractor

FIGURE 9-7   Inside a LiFi plasma luminaire. *(Pemco Lighting)*

Plasma lights are still a niche market, but boosters believe that they may become more versatile and cost competitive in the near future.

High-Concept Future Systems

There are many ways lighting could be improved in the future. One innovative design was submitted recently by Clay Moulton as part of his master's thesis at Virginia Tech's College of Architecture and Urban Studies. His Gravia concept is a LED floor lamp that combines a bit of human power and the action of gravity to produce 600 to 800 lumens (roughly equivalent to a 40-watt incandescent bulb) without ever being plugged in.

The lamp has a weight that can be raised manually, say, every four hours. As the weight slowly descends, the energy is converted to torque via a high-efficiency ball screw. This drives gears that spin magnets and power the LEDs. Obviously, such a device won't be for everyone, but it does hint at the essentially limitless opportunities if we think outside the (light)box.

Summary

Emerging green lighting technologies offer considerable promise to decrease energy use further, increase sustainability, and bring illumination to millions of people around the world who currently lack it. Solar lighting also can offer flexibility for certain applications in the developed world. Next-generation lighting technologies are receiving considerable investment and soon may offer even greater promise.

So what have we learned? We have seen that there are many ways to green up our lighting, whether we live in a private home or manage a large commercial facility. We can take advantage of more efficient technologies such as fluorescents, LEDs, and more, and pair these with better fixtures, sensors, and controls. We can learn to light everything smarter, increasing comfort, beauty, and productivity while trimming waste. We now have some options for lighting hardware with recycled and nontoxic content, and more should be coming online soon. We can try to reduce environmental impact through limiting packaging and transportation.

For green lighting, the future is very bright. And by making a few relatively simple changes now, you can start seeing the benefits immediately.

Resources

Government

U.S. EPA Energy Star Hotline
1200 Pennsylvania Ave. NW
Washington, DC 20460
(888) STAR-YES (888-782-7937)
www.energystar.gov

**U.S. Department of Energy's Office of Energy Efficiency
and Renewable Energy (EERE)**
877-EERE-INF (877-337-3463)
www.eere.energy.gov

Database of State Incentives for Renewable Energy (DSIRE)
Funded by the DOE, DSIRE is a great source of information on state,
local, utility, and federal incentives on renewable energy and energy
efficiency.
www.dsireusa.org

Office of Energy Efficiency (Canada)
580 Booth Street
Ottawa, ON K1A 0E4, Canada
613-996-4397
http://oee.nrcan.gc.ca/english

Tools

ENERGYGuide.com

A handy site that connects users with local information on green lighting and other energy resources.
16 Laurel Avenue, Suite 100
Wellesley Hills, MA 02481
781-694-3300
info@energyguide.com
www.energyguide.com

Green Made Simple

Videos and other content on how to green up home energy use, including up-to-date information on rebates and tax credits.
www.greenmadesimple.com

The Home Energy Saver (via Energy Star)

A user-friendly, Web-based do-it-yourself (DIY) energy-audit calculator, visited by 1 million people a year. Makes it easy to find ways to save in your home.
http://hes.lbl.gov

LampRecycle.org

Information on recycling spent mercury-containing lamps.

Advocacy

Add Your Light

Nonprofit bringing solar lighting to Haiti and the Dominican Republic.
1506, 80 Pt. McKay Circle NW
Calgary, Alberta T3B 4W4, Canada
403-670-0162
www.addyourlight.org

Alliance to Save Energy

A nonprofit coalition of business, government, environmental, and consumer interests that does research, advocates policy, forms public-private partnerships, and leads a number of educational programs, including the popular Energy Hog efficiency program (www.energyhog.org).

1850 M Street NW, Suite 600
Washington, DC 20036
202-857-0666
info@ase.org
http://ase.org

American Council for an Energy-Efficient Economy (ACEEE)

Nonprofit organization dedicated to advancing energy efficiency.

529 14th Street NW, Suite 600
Washington, DC 20045-1000
202-507-4000
ace3info@aceee.org
www.aceee.org

D.Light Design

Company dedicated to supplying low-cost solar lighting to those in need in developing countries.

360 Bryant Street, Suite 100
Palo Alto, CA 94301
www.dlightdesign.com

ecoDrinking

Website dedicated to promoting eco-friendly bars, clubs and lounges, with a strong emphasis on greening the lighting of these establishments.

http://ecodrinking.com

Environmental Working Group

A leading environmental research and advocacy group that tests consumer products, including lighting, for safety, quality, and green credentials.

1436 U Street NW, Suite 100
Washington, DC 20009
202-667-6982
www.ewg.org

Illuminating Engineering Society of North America
Industry association working to improve lighting.
120 Wall Street, 17th floor
New York, NY 10005
212-248-5000
www.ies.org

Rock The Reactors
Advocacy group dedicated to promoting green lighting, especially
light-emitting diodes (LEDs), in part to end the need for nuclear
power.
25 Newtown Turnpike
Weston, CT 06883
www.rockthereactors.com

Rocky Mountain Institute
A progressive energy think tank that has been at the forefront of renewable energy, green building, and efficiency technologies.
2317 Snowmass Creek Road
Snowmass, CO 81654
970-927-3851
www.rmi.org

SunNight Solar (BoGoLights)
For every solar flashlight sold in the developed world, a unit is
shipped to someone who needs one in the developing world.
5802 Val Verde Street, Suite 100
Houston, TX 77057
www.bogolight.com

U.S. Green Building Council (USGBC)
The community of professionals—organized into regional chap-
ters—working to spread green building by serving as a resource, ad-
ministering the LEED certification program, and advocating policy.
2101 L Street NW, Suite 500
Washington, DC 20037
www.usgbc.org

Retailers

1000Bulbs.com
One of the Web's most popular destinations for every kind of lighting product; offers discount prices.
2140 Merritt Drive
Garland, TX 75041
800-624-4488
www.1000bulbs.com

Bambeco
Online purveyor of gorgeous green products for the home, including solar and other unique green lighting.
P.O. Box 661
Moorefield, WV 26836
866-535-4144
www.bambeco.com

Ecohaus
A green superstore in Seattle, Portland, and San Francisco.
help@ecohaus.com
www.ecohaus.com

Environmental Lights
Online retailer of green lighting products.
11235 West Bernardo Court, Suite 102
San Diego, CA 92127
888-880-1880
www.environmentallights.com

Gaiam
One of the biggest names in green living, Gaiam offers many products in print catalogs and online.
833 W South Boulder Road
P.O. Box 3095
Boulder, CO 80307-3095
877-989-6321
www.gaiam.com

GREENCulture

Online retailer with a large and high-quality selection of eco-friendly lighting products.
32 Rancho Circle
Lake Forest, CA 92630
877-20-GREEN
www.eco-lights.com

Green Depot

The Green Depot started as a source of green building materials for contractors and has expanded to 10 stores in the Northeast and Chicago, plus robust Web ordering for individuals and commercial customers. Visit the gorgeous LEED Platinum-certified flagship store in Manhattan.
222 Bowery
New York, NY 10012
877-883-4733
contactus@greendepot.com
www.greendepot.com

Green Home

This Web-only store is a good source of green lighting products.
850 24th Avenue
San Francisco, CA 94121
www.greenhome.com

GREENandSAVE

Offers lease-to-own and savings-sharing LED programs that require no money up front from customers.
204 Old Lancaster Road
Devon, PA 19333
info@ledsavingsolutions.com
www.ledsavingsolutions.com

Greener Country

A retailer with a wide array of green products, available online and in the store on Long Island.
457 N. Broadway
Jericho, NY 11753
877-680-6010
info@greenercountry.com
www.greenercountry.com

My Solar Shop
Online retailer of solar lights.
C/O Protech Software Solutions LLC
P.O. Box 3598
Paducah, KY 42003-3598
270-443-5234
www.mysolarshop.com

OutdoorSolarStore.com
Carries a wide range of outdoor solar lights.
800-985-4129
www.outdoorsolarstore.com

Real Goods
One of the original sources of green products via catalog and online,
offering dozens of lighting options.
833 W. South Boulder Road
Louisville, CO 80027
800-919-2400
www.realgoods.com

The Ultimate Green Store
Online marketplace for green products, including lighting.
800-983-8393
www.theultimategreenstore.com

Lighting Manufacturers

BlueMax/Full Spectrum Solutions
Offers an array of light therapy and general compact fluorescent lighting (CFL) products with full-spectrum and high-definition quality.
P.O. Box 1087
Jackson, MI 49204
866-366-4029
www.bluemaxlighting.com

ClearLite/TAG Industries

Makers of the ArmorLite safety CFL and other green lighting products.
102 NE 2nd Street, Suite 400
Boca Raton, FL 33432-3908
800-514-5500
Sales@ClearLite.com
www.clearlite.com

Feit Electric

Manufactures a full line of lighting products from low-mercury CFLs to halogens, LEDs, and more.
4901 Gregg Road
Pico Rivera, CA 90660-2108
562-463-BULB (2852)
info@feit.com
www.feit.com

Fire & Water

David Bergman's boutique line of green lighting fixtures using recycled and nontoxic materials and powered by fluorescents and LEDs.
241 Eldridge Street, 3R
New York, NY 10002
212-475-3106
www.cyberg.com/fw/fw.htm

GE Lighting

Descended from Thomas Edison's company, GE is one of the best-known makers of incandescents, fluorescents, halogens, LEDs, and more, with major research and development efforts on advanced lighting underway.
800-GELIGHT
www.gelighting.com

Home Automation Inc.

Manufacturer of lighting, energy, and electronics control products for homes and businesses.
4330 Michoud Blvd.
New Orleans, LA 70129
(504) 253-2958
www.homeauto.com

Litetronics
Makers of high-quality CFLs (including the low-mercury EarthMate line), cold-cathode CFLs, halogens, high-intensity discharge lights (HIDs), and fluorescents.
101 W. 123rd Street
Alsip, IL 60803
708-389-8000
CS@Litetronics.com
www.litetronics.com

Lutron
The world leader in lighting controls.
7200 Suter Road
Coopersburg, PA 18036-1299
888-LUTRON1
www.lutron.com

Luxim
Pioneers of "light-emitting plasma" lighting.
1171 Borregas Avenue
Sunnyvale, CA 94089
408-734-1096
www.luxim.com

MaxLite
Maker of low-mercury CFLs and other efficient lighting systems.
80 Little Falls Road
Fairfield, NJ 07004
800-555-5629
info@maxlite.com
www.maxlite.com

OkSolar
Purveyors of solar, LED, and other advanced lighting products.
11956 Miramar Parkway
Miramar, FL 33025
347-624-5693, ext: 2
SalesUS@OkSolar.com
www.oksolar.com

Osram Sylvania

A world leader in lighting products, especially CFLs and fluorescents.
100 Endicott Street
Danvers, MA 01923
800-LIGHTBULB (1-800-544-4828)
www.sylvania.com

Pemco Lighting

Makers of plasma luminaires and other lighting products.
150 Pemco Way
Wilmington, DE 19804
302-892-9000
www.pemcolighting.com

Philips

Another leading lighting manufacturer, offering high-quality CFLs, fluorescents, LEDs, and more.
www.lighting.philips.com
Philips Lumileds (LED Division)
370 West Trimble Road
San Jose, CA 95131
408-964-2900
info@lumileds.com
www.philipslumileds.com

Plasma International

Focused on developing, producing, and commercializing new Sulphur Plasma Lighting Systems.
www.plasma-i.com

Sea Gull Lighting

Offers a range of efficient lighting products.
301 West Washington Street
Riverside, NJ 08075-4142
800-347-5483
info@seagulllighting.com
www.seagulllighting.com

Sol, Inc.
Manufacturer of solar-powered lighting since 1990.
3210 S.W. 42nd Avenue
Palm City, FL 34990
772-286-9461
www.solarlighting.com

LED Specialists

Aeon Lighting Technology
3086 Balmoral Drive
San Jose, CA 95132
info@aeonlighting.com
www.aeonlighting.com

American Green Energy Council
7075 Treadway Road
Port Charlotte, FL 33981
941-268-1653
www.atgledlighting.com

For a state-by-state list of commercial electrical contractors who are familiar with green lighting, check out www.atgledlighting.com/installers_ep_38-1.html

American Lighting
7660 East Jewell Avenue, Suite A
Denver, CO 80231
303-695-3019
www.americanlighting.com

ANL Lighting
188 Washington Street
Poughkeepsie, NY 12601
845-635-1121
info@andyneallighting.com
www.andyneallighting.com

BetaLED
1200 92nd Street
Sturtevant, WI 53177-1854
800-236-6800
sales@betaled.com
www.betaled.com

Boca Flasher
552 NW 77th Street
Boca Raton, FL 33487
561-989-5338
www.bocaflasher.com

Brilliance LED
Xin Mu Industrial Park, Building 36, 3rd floor
Xin Mu Cun, Ping Hu Town, Longgang District
Shenzhen, People's Republic of China 518111
+86-755-6120-1818
sales@brilliance-tech.com
www.brilliance-led.com

CREE
4600 Silicon Drive
Durham, NC 27703
919-313-5300
LEDlampSales@cree.com
www.cree.com

Dialight
1501 Route 34 South
Farmingdale, NJ 07727
www.dialight.com

EarthLED
Advanced Lumonics, LLC
7491 N Federal Highway, C5-251
Boca Raton, FL 33487
561-997-2509
EarthLEDSales@EarthLED.com
www.earthled.com

ElektoLed, Ltd
Unit 2A
Red Lions Business Centre
Burnham Road
Latchingdon Essex CM3 6EY, United Kingdom
+44 (0) 1621 741460
sales@elektoled.com
www.elektoled.com

Fawoo Technology
739-8, Ojeong-dong, Ojeong-gu
Bucheon, Gyeonggi-do
Republic of Korea 421-170
+82-32-670-3000
www.fawoo.com

GBL LED Lighting
209–460 Nanaimo Street
Vancouver, BC V5L 4W3, Canada
604-216-0448
sales@gbl-led.com
www.gbl-led.com

ILUMINARC
3000 N. 29th Street
Hollywood, FL 33020
954-923-3680
support@iluminarc.com
www.iluminarc.com

Infinilux
1457 Glenn Curtiss Street
Carson, CA 90746
sales@infinilux.com
www.infinilux.com

Juno Lighting Group
800-367-5866
www.junolightinggroup.com

LEDdynamics
Suite 100
44 Hull Street
Randolph, VT 05060
802-728-4533
everled_sales@leddynamics.com
www.everled.com

LEDinside
4F, No. 68, Sec. 3
Nanjing East Road
Taipei, 104, Taiwan
415-742-8633 (US)
+886-2-7702-6888 (TW)
service@ledinside.com
www.ledinside.com

LEDtronics
23105 Kashiwa Court
Torrance, CA 90505
310-534-1505
www.ledtronics.com

LED Holiday Lighting
Seasonal Impressions
864 Bridle Creek Drive
Jordan, MN 55352
888-657-6510
www.ledholidaylighting.com

LED Linear North America
121 Logan Avenue
Toronto, ON M4M 2M9, Canada
416-356-4461
www.led-linear.com

LED Online
P.O. Box 4145
Bath BA1 0GJ, United Kingdom
020 81238238
info@ledonline.co.uk
www.ledonline.co.uk

LED Waves
33 35th Street, 6th floor
Brooklyn, NY 11232
347-416-6182
customerservice@ledwaves.com
www.ledwaves.com

Lemnis Lighting (Pharox LED)
512 Second Street, 4th floor
San Francisco, CA 94107
888-7-PHAROX
info@myPHAROX.com
www.mypharox.com
www.lemnislighting.com

Light Emitting Designs
108 S. Wynstone Park Drive, Suite 103
North Barrington, IL 60010
847-380-3540
sales@led-llc.com
www.led-llc.com

Lighting & Electronic Design
141 Cassia Way, Suite C
Henderson, NV 89014
800-700-5483
mail@ledlinc.com
www.ledlinc.com

Line Lite International BV
Oranje Nassaulaan 58
1075 AS Amsterdam, The Netherlands
+31 (0)20 - 664 22 81
info@linelite.com
www.linelite.com

National LED Direct Corporation
550 Alden Road, Unit 113
Markham, ON L3R 6A8 Canada
416-934-5566
info@nationalleddirect.com
www.nationalleddirect.com

OPTILED Lighting International, Ltd.
Suite 2302, 23/F, One Landmark East
100 How Ming Street
Kwun Tong, Hong Kong
+852 2607 4268 / 3961 6000
USA: 11 Broadcommon Road, Suite 3
Bristol, RI 0289
401-396-9640
www.optiled.com

Optilume
2419 Ellwood Drive SW
Edmonton, AB T6X 0J6, Canada
780-432-5900
www.optilume.net

ROHM Semiconductor USA
10145 Pacific Heights Boulevard, Suite 1000
San Diego, CA 92121
858-625-3600
www.rohmelectronics.com

ShenZhen Bang-Bell Electronics
JinXiongDa Industrial Park, Building C
Huan'guan South Road
Guanlan Town, Bao'an District
Shenzhen, Guangdong, People's Republic of China
+86 138 2317 7749
www.bbeled.com

Sunovia Energy Technologies
6408 Parkland Drive, Suite 104
Sarasota, FL 34243
941-751-6800
sales@sunoviaenergy.com
http://sunoviaenergy.com

Super Bright LEDs
4400 Earth City Expressway
St. Louis, MO 63045-1328
314-972-6200
www.superbrightleds.com

Universal Display Corporation
Specializes in OLEDs.
375 Phillips Boulevard
Ewing, NJ 08618
609-671-0980
www.universaldisplay.com

Magazines

Home Power Magazine
Subscription Services
P.O. Box 520
Ashland, OR 97520
541-512-0201
http://homepower.com

LED Journal
7355 E. Orchard Road, Suite 100
Greenwood Village, CO 80111
720-528-3770
www.ledjournal.com

LED Magazine
PennWell International Publications, Ltd.
16 Arlington Villas
Bristol BS8 2EG, United Kingdom
leds@pennwell.com
www.ledsmagazine.com

Books

Green Living: The E Magazine Handbook for Living Lightly on the Earth
    Plume, 2005

Healthy House Building: A Design & Construction Guide. Healthy
  House Institute, 1997

Whole Green Catalog
Rodale Press, 2009
http://wholegreencatalog.com

Note: Special thanks to Remy C. of Rock The Reactors for helping to
put this resource list together.

Index

Environmental Lights.com

Your Resource for LED Lighting.

When it comes to lighting, EnvironmentalLights.com is your one-stop resource for high quality, energy-efficient lighting and LED products. We search the globe for best-in-class sustainable lighting and controls, and extensively test our products to meet the needs of your commercial or residential projects. And, because we are experts at understanding the specific needs of each of our clients, our technicians offer unparalleled support and expert advice. EnvironmentalLights.com. It's a new angle on green.

www.EnvironmentalLights.com | 888.880.1880

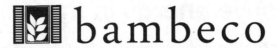
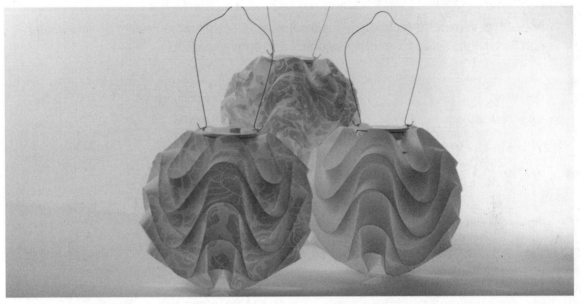

Lutron®—save energy in the perfect light

Maestro Wireless® offers simple, energy-saving light control solutions for any room in your home.

Maestro Wireless technology makes it easy to retrofit projects with no new wiring and has simple setup for effortless control and convenience.

Dimmers

- Start with a Maestro Wireless dimmer for elegant, intuitive control of a light or group of lights

dim lights	electricity saved
10%	10%
25%	20%
50%	40%
75%	60%

Occupancy Sensors

- Combine it with Radio Powr Savr™ wireless occupancy/vacancy sensor to automatically turn lights on when a room is occupied and off when it is vacant
- One sensor can control up to 10 dimmers or switches

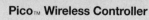

Potential Energy Savings

20%

Plug-in Lamp Module

- Add a plug-in lamp module to integrate table and floor lamps into the system for additional light control

dim lights	electricity saved
10%	10%
25%	20%
50%	40%
75%	60%

Pico™ Wireless Controller

- Complete the system with a Pico to provide convenient light control from anywhere in the space
- Pico can control up to 10 dimmers or switches

Stand   Wall mount   Hand held

For more information on **Maestro Wireless** contact your local Electrical Distributor, call **1.877.2LUTRON** or visit **www.lutron.com/maestrowireless**.

©2010 Lutron Electronics Co., Inc.

save
energy
with
Lutron™

The Case for Green

It's not hard to make the bottom-line case for green buildings. But studies are making deeper connections between green buildings, productivity, and occupant well-being. Consider the following claims.

For businesses:
- In a Reno post office, the quality of lighting cut sorting errors to 0.1 percent, the lowest in the region.[4]
- The productivity gains paid for the renovation in less than a year.[4]

For schools:
- Good lighting may improve test scores.[5]
- And plays a significant role in the achievement of students.[5]

For healthcare setting:
- Daylight and optimized artificial lighting can potentially reduce staff error rates and increase staff retention.[6]
- And potentially hastens patient recovery.[6]

Philips believes that living, learning, or working in a green building nurtures a culture in which people become more committed to sustainability.

It's 21st-century proof of Winston Churchill's astute observation: "We shape our buildings; thereafter they shape us."[7]

FACT:

Philips Energy Saving CFLs allow simple, screw-in replacement of incandescent lamps as well as maximizing energy savings compared to incandescents.

4) Joseph J. Romm and William D. Browning, "Greening the Building and the bottom Line: Increasing Productivity Through Energy-Efficient Design," Rocky Mountain Institute, 1998, p. 4
5) www.cap-e.com, A Capital E Report, "Greening America's Schools. Costs and Benefits." October 2006, p.11.(Citing Turner Construction, 2005 Survey of Green Buildings.)
6) www.cap-e.com, "Lighting Efficiency: Healthcare-Top 5 Green Building Strategies." EPA Publication 909-F-07-001, p.1
7) Leonard Roy Frank, "Webster's Quotationary." Random House. edition 2001, p. 39

REDUCE. REUSE. RECYCLE.

Reducing energy costs and waste is easy with efficient, long-lasting lighting products from Philips. But your commitment to sustainability doesn't end there—and neither does ours.

For a "cradle-to-cradle" green solution, Philips has partnered with Earth Protection Services, Inc. (EPSI) to help customers recycle lamps and ballasts. Using simple online ordering, Philips customers can use EPSI-PAK to help "close the loop" on lighting.

Please visit : http://www.nam.lighting.philips.com/us/recycle/ for additional information on recycling.

Philips Compact Fluorescent Lamps: Leading in quality and choice

Energy Saving CFLi Lamps

Can a simple switch of light bulbs save the world? The EPA's ENERGY STAR® website states that "if every American home replaced just one light bulb with an ENERGY STAR–qualified bulb, we would save enough energy to light more than 3 million homes for a year, more than $600 million in annual energy costs, and prevent greenhouse gases equivalent to the emissions of more than 800,000 cars."[10]

Philips is a leader in CFL technology, and our compact fluorescent lamps provide a direct replacement for incandescent lamps, delivering an incandescent-like light, too. And our range of choices means a replacement for most fixtures, regardless of shape, wattage, lumen output, color temperature or dimming options desired.

Energy Saving CFLni Lamps

For a green choice with maximum energy savings, switch out PL-L 40W lamps with Philips Energy Advantage PL-L 25W lamps—and slash energy use by 20%.[11] Plus you'll reduce mercury content, thanks to Philips ALTO® technology.

The compact design of the Energy Advantage lamps offers light output comparable to a 25W 4-foot linear fluorescent, with 95% lumen maintenance and a wide range of color temperatures.

Philips continues to expand its offering of compact fluorescent non-integrated (CFLni) lamps. Look for energy-saving, lower wattage PL-C and PL-T products—more Sustainable Lighting Solutions from the original innovator in CFLni technology.

SAVING ON ENERGY COSTS IS AS EASY AS

CHANGING A LIGHT BULB

10) www.energystar.gov, "Compact Fluorescent Light Bulbs".
11) Using instant-start ballasts, a standard PL-L 40W lamp only draws 32 watts. For that reason, each Philips Energy Advantage PL-L 25W lamp saves 7 watts using a (32W - 25W = 7W).